PRAISE FOR
How to Teach Relativity to Your Dog

"For the price of a book, Orzel delivers the heady, joyful experience of taking a small college class with a brilliant and funny professor who really knows how to teach. A thoroughly winning romp through a rock-solid presentation of a beautiful subject."

—Louisa Gilder, author of *The Age of Entanglement*

"Move over, Krypto—there's a new superdog in town! Chad Orzel's dog Emmy, having mastered quantum physics, now helps us understand Einstein's theories of relativity in a deep and accessible way. Get this dog a cape!"

—James Kakalios, Professor of Physics, University of Minnesota, and author of *The Physics of Superheroes* and *The Amazing Story of Quantum Mechanics*

"Everyone's favorite physics-loving canine is back, this time giving us a dog's eye view of Einstein and relativity. Physics professor Chad Orzel leads Emmy (and us) through an engaging tour of light speed, time dilation, and amazing shrinking bunnies (length contraction)—not to mention what all this means for the search for the elusive 'bacon boson.'"

—Jennifer Ouellette, author of *The Calculus Diaries*

"With Nero, the egocentric cat who believes it is the center of the universe, and Emmy, the student dog whose questions and misunderstandings would drive any teacher to distraction, and whose interest in relativity is how $E = mc^2$ can turn squirrels into energy, Chad Orzel has created a delightful cast of characters to make his introduction to relativity relatively painless. A cleverly crafted and beautifully explained narrative that guides readers carefully into the depths of relativity. Whether you are a hare or a tortoise, or even a dog, you will enjoy this."

—Frank Close, author of *The Infinity Puzzle*

"Emmy may be one smart dog, but her owner also happens to be an uncommonly gifted communicator. Chad Orzel's treatment of special and general relativity is comprehensive, informative, and amazingly accessible, yet it's funny too. This is, by far, the most entertaining discussion of the subject that I've ever had the pleasure of reading."

—Steve Nadis, coauthor of *The Shape of Inner Space*

HOW TO TEACH RELATIVITY TO YOUR DOG

CHAD ORZEL

BASIC BOOKS
A Member of the Perseus Books Group
New York

Published by Basic Books,
A Member of the Perseus Books Group

Books published by Basic Books are available at special discounts for bulk purchases in the United States by corporations, institutions, and other organizations. For more information, please contact the Special Markets Department at the Perseus Books Group, 2300 Chestnut Street, Suite 200, Philadelphia, PA 19103, or call (800) 810-4145, ext. 5000, or e-mail special.markets@perseusbooks.com.

Designed by Pauline Brown
Typeset in 10.95 point Dante MT Std by the Perseus Books Group

Library of Congress Cataloging-in-Publication Data
Orzel, Chad.
How to teach relativity to your dog / Chad Orzel.
p. cm.
Includes bibliographical references and index.
ISBN 978-0-465-02331-8 (pbk.)—ISBN 978-0-465-02937-2 (e-book)
1. Relativity (Physics)—Humor. 2. Relativity (Physics)—Popular works.
I. Title.
QC173.57.O79 2012
530.11—dc23
2011044067

10 9 8 7 6 5 4 3 2 1

For Claire.
Did I type today?
Yes I did, honey.
I typed a lot.

CONTENTS

INTRODUCTION

IT'S COLD AND FLU SEASON AGAIN, and between teaching at the college and a toddler in day care, I get every single bug that goes around. I'm sitting at the dining room table grading exams when a coughing fit hits. When it finally stops, I take a drink of water, and then notice a thumping sound. I look over toward Emmy, on the floor next to the couch, and she's thumping her tail on the floor with her tongue lolling out the side of her mouth: the dog equivalent of a laugh.

"Yeah, laugh it up, fuzzball. You think this is funny?"

"Sorry," she says, "but at the end there, it sounded like you said—." She barks twice, sounding a little like a cough. "That's really funny in Dog."

"Yeah? What's it mean?"

"Well, it's . . . umm . . . You know, if you can't sniff your own butt, you won't get the joke."

"I'll try to contain my disappointment, then." I turn back to my grading.

"I'll think about it, and see what I can come up with, but translating humor is really hard."

"Translating anything is hard." I say, not looking up.

"Yeah? What do you know about translation?"

"Well, it's what I do for a living."

"You're a physicist, not a diplomat."

"I'm a physics *professor*," I say, putting my pen down. "In addition to doing physics, I teach physics to other people."

"And dogs!"

"Yes, and dogs. Teaching physics necessarily involves translation. The natural way to express physics is through math, but most people don't think in mathematical terms. So, a lot of the business of teaching physics is finding ways to translate physical ideas from mathematical equations into concepts drawing on everyday language and experience."

"So, making analogies and stuff like that?"

"That's part of it, yes. I also spend a lot of time dealing with people's preconceptions about how the world works. Sometimes, our intuition about how everyday objects behave leads us astray when we think about physics, and the first step in teaching the subject is to break down those preconceptions. Basically, to start over."

"You wouldn't have that problem if you stuck to teaching dogs," she says, looking pleased with herself and her species.

"No?"

"Nope. I have no clue at all about how things work. I'm a clean slate, when it comes to physics."

"I wouldn't go that far, but you at least have a *different* set of preconceptions than most humans do. Which means that thinking about physics as it appears to a dog can be a useful thing to do—looking at the problem from a different angle, and with an open mind, can sometimes give you insight that you wouldn't get by going straight at your own misconceptions."

"So, when you think about it, teaching physics to me helps you teach physics to humans."

"Yeah, it does."

"Which means that in a sense, it's part of your job, right?" She trots over to me and sits down, looking hopeful.

"I know where you're going with this, so let me remind you that grading these papers is *also* part of my job. I need to turn my final grades in tomorrow, so that's the more important part right this minute."

"Oh." She deflates a little.

"But tomorrow is also the start of our break, so I'll have time to spend talking to you about physics, if you want."

"Preferably while taking long walks!"

"Sure, that works. So, let me finish grading these exams, while you think about what areas of physics you'd like to learn about, OK?"

"OK!" She trots off in the direction of the library, and I go back to my grading. As I start on the next paper, I hear her saying "Maybe I can finally find out what this Einstein guy was all about . . ."

Ask any human, or most dogs, to picture a scientist, and odds are good that their mental image will look a lot like the iconic pictures of Albert Einstein—white hair sticking in all directions, rumpled clothes, maybe even a German accent and a distracted air. This is a little unfair to scientists*—great scientists come in all sizes, shapes, races, genders, and nationalities (though not yet species)—but Einstein has captured the popular imagination to an amazing degree and dominates the popular image of a scientist. Even more than fifty years after his death, Einstein was the second most popular answer in a poll asking people to name a living scientist**.

---

* The image is also a little unfair to Einstein himself, who was not even thirty when he published his revolutionary work. Pictures of him then show a handsome and well-groomed young man—the wild hair and baggy clothes came much later.

** Among those who could name a scientist at all, that is—65 percent couldn't even make the attempt (http://newvoicesforresearch.blogspot.com/2009/07/can-you-name-living-scientist.html). Stephen "A Brief History of Time" Hawking had the top spot.

Asked *why* Einstein is a famous scientist, even dogs can come up with the equation $E=mc^2$, and possibly the words "theory of relativity." Explaining what those *mean*, and where they come from, is beyond most humans, though, let alone dogs. This is an unfortunate state of affairs, as Einstein's theory of relativity is one of the cornerstones of modern physics. Along with quantum mechanics, relativity completely revolutionized the way scientists view our universe. It provides insight into problems that classical physics can't handle, and poses new problems that physicists still grapple with a hundred years later.

Unfortunately, the features that make relativity so essential to physics also make it extremely intimidating to non-physicists. Relativity deals with situations that are very foreign to our everyday experience of the universe—objects moving thousands of times faster than the fastest man-made objects, astronomical objects packing enormous masses into tiny spaces—and its predictions defy all our normal expectations. Relativity tells us that quantities that seem fundamental—distances through space, and duration in time—in fact vary from one observer to another. A moving clock ticks at a different rate than a stationary one. A clock near a massive object ticks at a different rate than one farther away. And space itself is stretched by the presence of mass, so the length of a path between two points depends on what you pass along the way.

These are all surprising predictions made by the theory of relativity. They are also unequivocally true, confirmed by countless experiments in the century since Einstein first introduced relativity in 1905. The universe we live in is a far stranger place than our everyday intuition leads us to expect. To fully understand it, we have to expand our conception of the universe to include the counterintuitive predictions of the theory of relativity.

This can seem a daunting task, but one way to make it more approachable is to think like a dog. As any pet owner knows, dogs look at the world in a very different way—not entirely without preconceptions, but at least with different preconceptions than their humans, often in ways that make physics easier to understand. To a dog, any time *should* be dinner time, so the idea of clocks running at different rates for moving observers, or observers in different places, is easier to accept.

If you can learn to think like a dog, to approach the world as an endless source of surprise and wonder, modern physics is much less intimidating. Looking at physics from a dog's point of view allows us to shake off some of our human expectations about how things ought to work, and lets us appreciate the weird and wonderful world revealed by relativity.

This book reproduces a series of conversations with my dog about aspects of both special and general relativity. Each conversation is followed by a more detailed discussion of the physics involved, aimed at interested human readers. Some of the topics covered have achieved fame, or at least notoriety, in the wider culture, like Einstein's famous equation $E=mc^2$ (Chapter 7) or the idea of black holes (Chapter 10); others are less familiar to non-physicists, such as the merging of space and time (Chapter 5) or the effects of gravity on time (Chapter 9), but are just as essential to the modern understanding of physics. We'll also talk about some of the innumerable experiments and observations confirming that the universe is a weird and wonderful place.

We won't be able to cover everything that's interesting about relativity—generations of scientists have dedicated their careers to the subject without managing to exhaust its wonders—but we hope this will provide an introduction to the subject giving human and canine readers some sense of what it's about, and why it's important. And the next time some pesky cat asks you to explain *why* Einstein is famous, you'll have a good answer for them.

"Don't forget me!"

"I'm not forgetting you. How am I forgetting you?"

"Oh, sure, you mention the conversations at the start of the chapters. But you didn't tell them that I'll be keeping an eye on you in the middles, too. If you try to leave anything out, or sneak something past without explanation, I'll make sure you get it right."

"You mean, like you're doing now?"

"Yeah. Oh, wait—is this bit going in the book?"

"Yes."

"Oh. Well, then, I guess they're informed. But they should know that I'll be keeping an eye on you, and I'm an *excellent* watchdog."

"I think you've made that point."

"Also, I'm really cute, and I like long walks, and belly rubs, and chasing bunnies, and bacon. I really like bacon. Also, cheese. And peanut butter."

"I fail to see how this is relevant."

"Well, you know, in case they want to mail me presents. You know, because I'm an excellent physics dog, and all. You are going to put our address in the book, right? So they can send me stuff."

"I think that's just about enough out of you."

"Oh, all right. You're no fun, though."

"Can we please get started with the physics discussions?"

"Sure, absolutely. Lay some physics on me—I'm ready for anything!"

Chapter 1

RELATIVE DOG MOTION: THE DESCRIPTION OF MOTION

As I'm driving down the street, a squirrel darts out into the road a block or so ahead of me. From the back seat, Emmy says, "Gun it! Hit the squirrel, hit the squirrel, *hitthesquirrel*!"

"Will you sit down and be quiet?" We're having some work done on the house, and I'm taking her to campus with me so she's not underfoot for the contractors.

The squirrel makes it to the other side of the road and up a tree to safety. "Awwww," Emmy says. "Dude, you totally could've gotten that one. This car is way faster than a stupid squirrel."

"That may be, but I have a class to teach. I don't have time to careen around like a maniac chasing squirrels with the car."

"No, no—you'd have plenty of time. Time slows down when you go faster."

I look in the rearview mirror. Emmy's standing on the seat, wagging her tail and looking pleased with herself.

"Oh, God," I say. "Don't tell me you've started reading about relativity."

"OK, I won't tell you." She's quiet for a few seconds, then, "Relativity is pretty cool, though. I can slow time!"

This is not going anywhere good, I can tell. We come to a traffic light, and I stop.

"One thing I don't understand, though . . ."

I sigh. "OK, what is it you want me to explain?"

"Why do they call it that?"

"Why do they call what, what?"

"Why do they call relativity 'relativity'? Why not something cooler, like superfast time-slowing squirrel-catching dynamics?"

"Well, for starters, physicists don't care much about squirrels. More importantly, though, the name 'relativity' comes from one of the theory's most basic elements: the idea that relative motion is the only thing that matters. There is no absolute **frame of reference** against which we can measure the motion of everything in the universe." The light changes, and I start driving again.

"Yeah, but that's silly. Of course there's a fixed frame of reference."

"Really? What is it?"

"Well, our house, silly. And the yard, with the big oak tree. And the other tree. And the other, other tree. And—"

"OK, OK, I get it."

"The house is where I keep my stuff!"

"Yeah, OK. But the house only *looks* like it's a fixed frame of reference. I mean, it's on the Earth, right? And the Earth is rotating."

"I guess so . . ."

"And it also moves around the sun, which is why we have seasons. The thing you're using as a fixed reference point is really in constant motion, and all you're doing is measuring your motion relative to it."

"OK, but I can still tell the difference between when I'm standing still and when I'm moving."

"How?"

"Well, when I'm moving, I walk past stuff, and sniff things, and chase bunnies and squirrels. When I'm not, I just sit there."

"Sure, but how can you tell the difference between a situation where you're moving, and a situation where you're sitting still and everything else is moving in the opposite direction?"

"Well, that would be silly." We come to another red light, and I stop again. "Anyway, I can tell that I'm the one moving, because my legs are moving."

"OK, but how about when you're in the car, like we are now?"

"What do you mean?"

"Well, we're sitting still right now, but when we start moving again . . ." The light changes, just at the right moment. I accelerate a bit, then cruise at a constant speed. "How can you tell that we're moving, rather than sitting still and watching the rest of the world move by?"

"Ummm . . . the engine is going."

"Yeah, but we could be on a treadmill, with fake scenery moving past us. This whole trip could be a fiendish illusion."

Emmy looks worried. "I don't like fiendish illusions."

"Calm down, it's just a hypothetical." She looks somewhat mollified. "Anyway, the answer is that there's no physics measurement you can do to distinguish between sitting still and moving at a constant **velocity,** the way we are now. You can detect **acceleration,** like this"—I step on the gas and speed up—"but when we're moving at a constant speed, all the laws of physics are exactly the same as when you're standing still."

"So how do you tell when you're moving?"

"You can't. All you can say is that you're moving *relative* to some other object—which is why the theory is called relativity."

I check the mirror, and she's looking thoughtful. "So," she says, "the only thing we can measure is relative velocity?"

"Exactly."

"Like your velocity relative to that car with the lights?"

"What?" I look behind us and see a police car pulling out, lights flashing. I look down and realize my foot is still on the gas. "Crap! Well, maybe he's after someone else . . ."

The cop car pulls in behind me. "I don't think so," Emmy says cheerfully. "He's got you nailed."

I pull over. "This is all your fault, you know," I say as I kill the engine.

"Yeah? Good luck explaining that to the cop." She turns toward the window and wags her tail cutely, just in case the police officer has dog treats.

Einstein's theory of relativity is one of the cornerstones of modern physics and requires a complete and dramatic rethinking of ordinary concepts of space and time. Its most famous predictions—the equivalence of mass and **energy,** the slowing of time for fast-moving observers, the warping of space near **black holes**—have captured the popular imagination and are staples of science fiction.

Relativity is exotic and exciting precisely because it runs so counter to our usual intuition—we don't notice its effects when going about our everyday lives. The effects of relativity only show up when we are dealing with either extremely small, fast-moving objects, like the subatomic particles being smashed together in **particle accelerators,** or extremely large, massive objects like black holes and galaxy clusters. It may seem surprising, then, that the place to start understanding the physics of relativity is with the motion of ordinary, everyday objects like dogs and cars.

NO MATTER WHERE YOU GO, THERE YOU ARE: DESCRIBING MOTION

While we associate the word "relativity" most strongly with Albert Einstein, the central idea goes back long before him. Einstein himself, in his popular book *Relativity: The Special and the General Theory,** attributes the

* First published in German in 1920. There are several English editions; mine is the 2006 Penguin Classics paperback.

concept to Galileo Galilei (1564–1642) and Sir Isaac Newton (1643–1727), who lived almost three centuries before he was even born.

The central idea of relativity is that the laws of physics must appear the same to all observers. Whether you're moving or standing still, physics should work in exactly the same way: objects should accelerate when pushed, energy and **momentum** should be conserved, and so on. This is just common sense—if the laws of physics were different for running dogs than dogs sitting calmly on the ground, it would be almost impossible to understand the world around us or to predict the motion of objects like bouncing balls or thrown treats. Since even dogs without a deep knowledge of physics can track down tennis balls and snatch treats out of the air, the simple version of relativity must be true.

Of course, human physicists aren't satisfied with "Watch me catch this tennis ball" as proof, so we need to be more precise about what we mean when we talk about moving objects. The key insights that led Einstein to the theory of relativity start from very careful and exact definitions of motion. Following in his footsteps, then, we need to talk about how physicists describe moving objects.

Emmy the canine physicist, sitting in her living room watching the world outside, can describe the positions of moving objects—Winthrop the basset hound walking by on the street, say, or a human child bicycling past—by measuring the distance from her spot by the window to the object in question. At some instant, Winthrop may be 9 meters (m) due south of her, for example. A child on a bike may be 8m south and 6m east.

Emmy can measure the motion of the creatures on the street by recording their positions at one time, waiting a bit, then looking at their new positions and comparing the two, as shown in Figure 1.1. If the small human on wheels moved from 8m south and 6m east to 8m due south in one second, our canine physicist would say that the child has a velocity of 6 meters per second (m/s) due west.* If Winthrop moves from 9m due south to 9m south and 3m east in the same time, he has a velocity of 3 m/s due east.

* It's very important to keep track of the direction of motion as well as the speed, as any dog who has ever tried to go up a down escalator can tell you.

Figure 1.1.

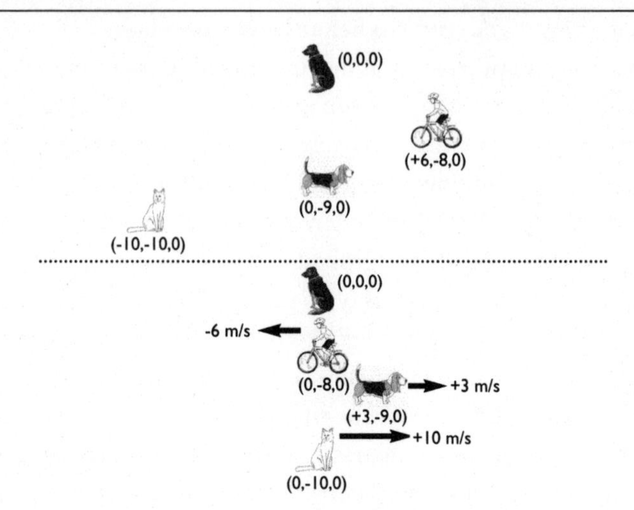

To start thinking about relativity, we need to ask how these objects appear to another observer—for example, my parents' yellow Lab, Bodie, in a car headed east along the street at, say, 10 m/s. That dog would measure the motion of the various other creatures by recording their distances from his own position in the car—

"No way, dude. Don't even go there."

"What?"

"You're about to say that dogs are very self-centered creatures and always measure positions relative to themselves. It's an unfair stereotype of dogs, and I won't stand for it."

"But that's how the whole thing works. If you want to understand relativity, you need to look at positions and velocities as measured by a moving observer."

"That may be, but that's not how dogs do it. We always measure positions relative to fixed points, like our houses."

"Look, I need a moving observer for this explanation to work out. Otherwise, this book is going to be a huge mess."

"You can have a moving observer, just don't make it a dog."

Figure 1.2.

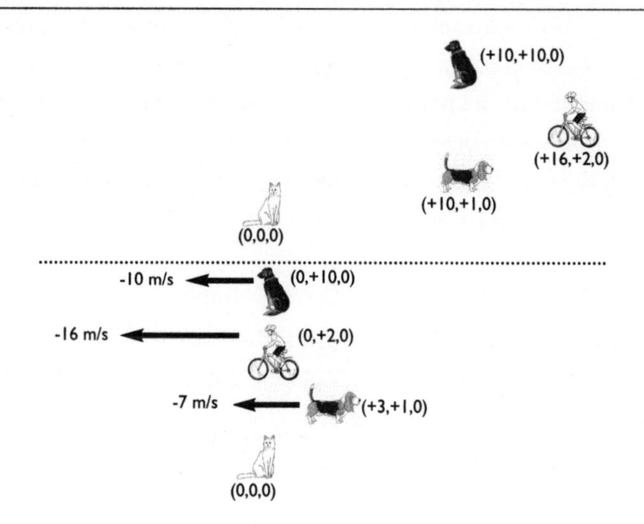

"Fine. How about a cat, then?"

"Oh, yeah, that's fine. Cats are *incredibly* self-centered."

To start thinking about relativity, we need to ask how these objects appear to another observer, for example, my sister's cat Nero riding in a car headed east along the street at, say, 10 m/s. The cat, being incredibly self-centered, would measure the motion of the various other creatures by recording their distances from his own position in the car. From Nero's point of view, shown in Figure 1.2, he is perfectly stationary, and the rest of the world moves around him. The barking dog at the window is actually moving west at 10 m/s, as are the house, the trees in the yard, and everything else in the world, according to Nero.

According to Nero in the car heading east, the westbound child on the bike is not moving at the 6 m/s measured by the stationary dog but at a higher speed. If the child is 20m east of the cat at some instant, one second later, the distance is down to just 4m: the child has pedaled her way west by 6m, while the road and everything on it has moved 10m west. The velocity of the bicycling child is now 16 m/s due west, or the speed of the car plus the speed of the bike.

Regarding Winthrop the eastbound hound, Emmy at the window and Nero in the car disagree about not only his speed but also his *direction* of motion. In the one-second interval between measurements, Winthrop has moved 3m east, but the road he is on has moved 10m west (according to Nero), so the distance between the two has actually decreased. According to Nero in the car, Winthrop is headed west at 7 m/s, rather than east at 3 m/s, as measured by Emmy in the window.

"I like this version much better. Cats are so dumb."

"He's got it right, though, at least in a mathematical sense."

"What do you mean? The house isn't moving west—that would be ridiculous."

"It seems ridiculous because we're used to thinking of the Earth as a fixed reference point and measuring motion relative to it. If you want to calculate what happens in interactions between Nero and some other object, though, it's much easier to work things out using his frame of reference."

"Frame of reference? Is the cat in a picture, now?"

"Sorry. Frame of reference is a physics term that refers to the measurements made by a particular observer. The dog in the window occupies a frame of reference in which she is stationary, and all distances are measured from her position. The cat in the car is in a different frame of reference, in which he is stationary, and all distances are measured from his position in the car."

"Yeah, but he's moving, and I'm not."

"In your frame, that's true. In his frame, though, you're moving, and he's standing still. The whole point of relativity is that these are equally valid ways of looking at the world. Both you and Nero can use physics to predict the motion of other objects, according to your own measurements, and get the right answers. And if you know how to convert measurements in one frame to measurements in another frame, you'll see that their answers are the same."

"So how do you convert from one frame to another?"

"I was getting to that when you interrupted me."

"Oh. Well, carry on, then."

The effect of changing frames is obvious for measured velocities, but observers moving at different speeds necessarily see different positions for things as well. Emmy sees Nero's position at one instant as 10m west of her window. One second later, he is exactly even with the window, and one second after that, he is 10m east of the window. Nero, on the other hand, sees Emmy start out 10m east of his position. One second later, she's even with the car, and one second after that, she's 10m west.

Physicists, whether canine or human, like to assign numbers to things, and position is measured by assigning three coordinates measuring the distance from some point of reference (Emmy's spot in the window or Nero's seat in the car) along each of three perpendicular directions—east-west, north-south, and up-down. We also know the time at which the measurements were made, so we can describe each using four numbers giving its "location" in space and time. The three space coordinates are traditionally referred to as x, y, and z, and time is t. We'll use x for the east-west direction, y for north-south, and z for up-down. To get from the reference point ($x = 0$, $y = 0$, $z = 0$) to the point ($x = 3$m, $y = 4$m, $z = 0$m), you would move 3m east, 4m north, and remain at ground level.*

To keep the numbers simple, let's call the point when Emmy the dog and Nero the cat have the same east-west coordinate time zero. In this case, Emmy would record Nero's coordinates as ($t = 0$; $x = 0$m, $y = -10$m, $z = 0$m). One second earlier, he was 10m west, so ($t = -1$s; $x = -10$m, $y = -10$m, $z = 0$m); one second later, he will be 10m east, so ($t = +1$s; $x = +10$m, $y = -10$m, $z = 0$m). From Emmy's point of view, her position remains constant at ($x = 0$, $y = 0$, $z = 0$) for all three times.

Nero, on the other hand, will record Emmy's position at these three moments as ($t = -1$s; $x = +10$m, $y = +10$m, $z = 0$m), ($t = 0$s; $x = 0$m, $y =$

---

* These numbers can be negative as well as positive, in which case the negative sign indicates a reversed direction: ($x = 3$m, $y = -4$m, $z = 0$m) is three meters east and four meters south of the reference point; −4m north is the same as +4m south.

+10m, $z = 0$m), and ($t = +1$s; $x = -10$m, $y = +10$m, $z = 0$m). The cat records his own position as ($x = 0$, $y = 0$, $z = 0$) for all three times. The space coordinates for times $t = -1$s and $t = 0$s are marked on Figures 1.1 and 1.2.

If you play around with these numbers a little, you can come up with a simple recipe for converting between Nero's measurements and Emmy's: you simply take the east-west coordinate measured by Nero, and subtract his speed (10 m/s) multiplied by the time. A little more fiddling around will show you that getting from Emmy's measurement to Nero's involves just the reverse: you take the east-west coordinate measured by Emmy, and add to it Nero's speed multiplied by the time. In this way, you can take any measurement made by Nero and convert it into a measurement that will make sense to Emmy, and vice versa.

We also see that the other coordinates don't change in time. Nero's measurement of the north-south position is always equal to Emmy's measurement plus 10m (the distance between her spot in the window and his car on the street), and they both agree that they're at ground level, not moving up or down. This is a very convenient feature of moving from one reference frame to another—coordinates that are perpendicular to the direction of motion do not change their values. For this reason, we will almost always deal with motion directed along one of the axes—that way, we don't have to worry about what happens to the other two position coordinates as we move from one frame to another.

The same recipe will work for converting any set of measurements made by one creature into another's frame of reference. You take the position measured by Nero and add his speed multiplied by the time to the x-coordinate; that gives you the x-coordinate measured by Emmy. Similarly, if you take the east-west velocity measured by Emmy and subtract Nero's velocity (keeping in mind that westbound velocities are represented by negative numbers), that gives you Nero's velocity measurement.

"Hey, dude? I thought relativity was all about moving really fast and catching squirrels and converting them into energy and stuff. When do we get to the squirrels?"

"We'll get there, but this background stuff is important. In relativity, even more than in other branches of physics, it's important to be extremely precise regarding what you're talking about. Otherwise, you get tied in knots."

"Yeah, but this stuff is sooooooo obvious. I mean, I see the cat moving, the moving cat thinks I'm moving, blah, blah, blah. I want to talk about Einstein!"

"Einstein's biggest contribution was showing that when you think extremely carefully about what it means to measure the motion of objects, relativity isn't weird at all. It's actually inevitable—relativity is the only possible way that the world could work."

"Yeah, but all this stuff about measuring positions . . ."

"The whole first section of Einstein's first paper on special relativity talks about how to synchronize two clocks at different locations."

"Oh."

"The second section talks about defining position via a giant three-dimensional grid of meter sticks filling the entire universe. His popular book does the same thing."

"Oh."

"So, cats in cars don't seem so bad, do they?"

"Cats are way more interesting than meter sticks, that's for sure. Carry on."

EVERYTHING HAS ALWAYS BEEN RELATIVE: ORIGINS OF RELATIVITY

We now know how to reconcile the different measurements of position and velocity made by observers who are moving relative to one another by adding or subtracting a factor that depends on the relative velocity. Given the position and velocity of an object as measured by one observer at some instant in time, we can determine the position and velocity measured by a different observer at the same time. The two sets of measurements will look different but are connected by a very simple operation.

Physics is about more than just describing motion, though—it's about predicting the future behavior of objects based on simple physical laws.

While it's nice to be able to reconcile measurements made by cats and dogs, what we really want is to use Emmy's measurements to predict what Nero will observe in the future, or vice versa.

To make this work requires the **principle of relativity:**

The laws of physics work the same way for an observer who is moving at constant velocity as for an observer who is standing still.

Two observers looking at the same events will both agree that the same laws of physics are in play, provided they are moving with constant speed relative to one another. They can each use the same laws of physics to predict the future position of a thrown object, or the operation of some mechanical device, once they know its position and velocity at some instant.

This might seem surprising, given that a moving object will have a different position and velocity according to each observer, but the world *has* to work this way. As we all know, a clever dog can catch a tennis ball thrown right to where she is sitting, and she can also run down a ball thrown into the yard. She can do both because the ball's motion is determined using the same physical laws in both frames. If the principle of relativity didn't work, a moving dog would need a different set of rules to predict the future position of a ball than a stationary dog, and a fast-moving dog would need a different set of rules than a slow-moving dog. Figuring out where the ball would land, and getting there before other dogs or humans, would be nearly impossible if the rules governing the motion of the ball changed depending on the speed of the dog.

This idea of relativity was introduced by the great Italian scientist Galileo Galilei, who used it in his *Dialogue Concerning the Two Chief World Systems*. The *Dialogue* was published in 1632 and consists of an imaginary conversation between three characters, one of whom, Salviati, tries to convince the other two of the validity of the Copernican theory that the Earth orbits the sun, rather than the other way around. Another character, Simplicio, puts forth various arguments against the Copernican view, all of which are deftly demolished by Salviati.*

One of Simplicio's arguments is that the Earth can't be moving around the sun, because it doesn't *feel* like we're moving. If we were moving, surely we'd know. Salviati disposes of this idea via a simple thought experiment:

> Shut yourself up with some friend in the main cabin below decks on some large ship, and have with you there some flies, butterflies, and other small flying animals. Have a large bowl of water with some fish in it; hang up a bottle that empties drop by drop into a wide vessel beneath it. With the ship standing still, observe carefully how the little animals fly with equal speed to all sides of the cabin. The fish swim indifferently in all directions; the drops fall into the vessel beneath; and, in throwing something to your friend, you need throw it no more strongly in one direction than another, the distances being equal; jumping with your feet together, you pass equal spaces in every direction. When you have observed all these things carefully (though doubtless when the ship is standing still everything must happen in this way), have the ship proceed with any speed you like, so long as the motion is uniform and not fluctuating this way and that. You will discover not the least change in all the effects named, nor could you tell from any of them whether the ship was moving or standing still. In jumping, you will pass on the floor the same spaces as before, nor will you make larger jumps toward the stern than toward the prow even though the ship is moving quite rapidly, despite the fact that during the time that you are in the air the floor under you will be going in a direction opposite to your jump.**

---

* The *Dialogue* is one of the earliest pieces of popular scientific writing. Galileo wrote it in Italian, rather than the Latin favored by scholars of his day, and it is a witty and engaging presentation of a scientific argument in terms understandable by a layman. It also got him in trouble with the Catholic Church, which was opposed to the Copernican model, and he spent his last years under house arrest. Since Galileo's day, most academic scientists have heeded his lesson and avoided writing for a popular audience.

** From the translation of the *Dialogue* by Stillman Drake (Berkeley: University of California Press, 1953), 186–187.

This passage introduces a common formulation of the principle of relativity: there is no experimental way to distinguish between a frame of reference that is standing still and one that is moving at constant speed. A cat riding on a cruise ship will, like Salviati's imaginary passenger, see all objects around him behaving in exactly the same way whether the ship is docked in a port or moving at speed through calm seas.

"Wait a minute—why is a cat on the cruise ship? I want to be on a cruise ship!"

"You objected when I wanted the moving observer to be a dog, so I made it a cat. Now you're stuck with cats as moving observers."

"Awwwww . . . I want to go on a cruise!"

"Those are the breaks, sorry."

"Anyway, it doesn't work."

"What do you mean?"

"Well, I can easily tell that I'm on a moving boat, because all the trees and stuff look like they're moving. Trees don't move, so if they look like they're moving, then I must be the one doing the moving."

"That's not an experiment on the boat, though. There is no test you can do *on the boat* that would indicate that you're moving rather than standing still. You can look outside the ship and apply logic, but remember, you could be duped by a fiendish illusion. Somebody could put a bunch of trees on rollers and make it look like you're moving when you're not."

"Oh, right. Stupid fiendish illusions."

In Galileo's day, the laws of physics hadn't been nailed down yet, so he wasn't able to do anything more quantitative than an appeal to common sense and experience. That had to wait for Isaac Newton, whose *Philosophiae Naturalis Principia Mathematica* (published in 1687)* is widely regarded as the work that first established physics as a mathematical science. In it, he set forth three laws of motion that govern the behavior of moving objects and form the core of "classical" physics.

---

* Newton, unlike Galileo, wrote his most famous work in highly technical Latin so as "to avoid being baited by little Smatterers in Mathematicks." This also gives you some idea of his personality.

Newton's first law is the principle of inertia: objects at rest tend to remain at rest, and objects in motion tend to remain in motion in a straight line at constant speed unless acted on by an external force. Newton's second law tells you how much force you need to change the speed of a given object and is often written as $F = ma$, or force equals mass times acceleration. Newton's third law says that for every action there is an equal and opposite reaction—a force of equal strength in the opposite direction.

With **Newton's laws** of motion in hand, we can make a concrete comparison between the laws of physics as seen by a stationary observer and the laws of physics seen by a moving observer. Let's consider the case of our two observers, Emmy and Nero, looking at the motion of a simple object, say, a ball thrown up into the air by a human standing next to the dog.

Emmy sees the ball fly straight up into the air, as shown in Figure 1.3. In accordance with Newton's laws, it doesn't move side to side at all, because there's no force acting on it in that direction. In the vertical direction, though, the force of **gravity** causes it to slow down, stop for an instant at the peak of its flight, then accelerate back down toward her. Meanwhile, the cat moves by from right to left at a constant speed, because there are no left-right forces acting on him.

So, what does Nero see? In Nero's frame of reference, shown in Figure 1.4, he is always stationary at $x = 0$, while the dog is moving from left to right. He sees the thrown ball move up and to the right, slowing gradually, then fall down and to the right, speeding up, tracing out a parabolic arc.

Nero's observations are also consistent with Newton's laws of motion. Emmy and the ball each move from left to right, and they each move the same amount. No force causes either the dog or the ball to change its motion in the horizontal direction, so they move at constant speed horizontally. The only force that acts is gravity, which only affects the vertical motion of the ball, causing it to slow down, reverse direction, then speed up, just as in Emmy's frame. At the very peak of the motion, Emmy sees the ball as completely stationary for an infinitesimal instant, while Nero sees the ball as moving from left to right with no up-down speed but the same left-right speed it's had all along.

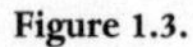

Figure 1.3.

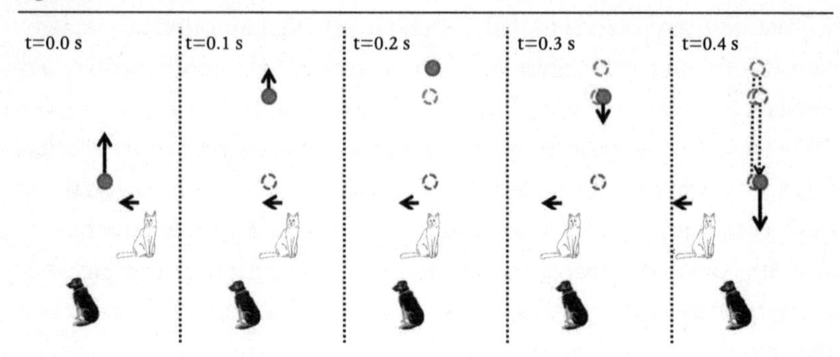

Figure 1.4.

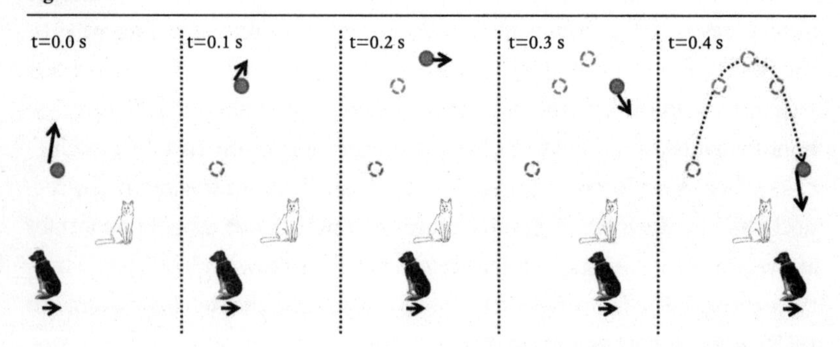

Cats and dogs don't agree on much, but this cat and this dog both agree that Newton's laws of motion govern the flight of the ball. The same would be true if Nero had thrown the ball. In his frame, he would see the ball go straight up and come straight back down, remaining directly above his head the whole time. Emmy, on the other hand, would see the ball trace out a parabolic arc, moving from right to left at the same speed as Nero, though she would agree that the ball is always directly above him.

"Yeah, yeah, yeah. Everybody always agrees that the laws of physics work. Get on with it."

"Actually, that's not true. Observers who are moving *at constant speed relative to one another* agree that the laws of physics work in the same way. There are other observers, though, who see something different."

"Like what?"

"Well, you remember that time I left a soda on the dashboard of the car?"

"Yeah. When you stepped on the gas, it fell off, and spilled all over the place. That was pretty funny. Also, yummy, because I got to lick soda off the seat."

"Well, that's an example of a situation where different observers would disagree about what happened. A stationary observer would say that the soda fell because there was no force acting on it, so in keeping with Newton's first law, it remained where it was while the dashboard accelerated out from under it."

"Sure, that makes sense."

"To an observer moving with the car, though, it looks like the soda spontaneously started moving by itself. A cat in the car would agree that nothing pushes the soda, but he would see it start accelerating toward the back of the car all the same, the minute I stepped on the gas. That observer would not agree that Newton's laws of motion applied to the cup of soda."

"So, if you're speeding up or slowing down, physics doesn't work?"

"Physics still works, but not as nicely as when you're dealing with motion at constant speed. Observers moving at constant speed relative to each other define a special group of **inertial frames,** so called because they are frames of reference in which Newton's principle of inertia (the first law) holds true."

"There are lots of other things you can do, though. Like, I can speed up, or slow down, or chase my tail around in a circle, or . . ."

"Exactly. We'll talk about some of that stuff later on, but for now, we're just going to deal with inertial frames, OK?"

"In a minute. First, I want to chase my tail for a bit. Wheee!"

"You're ridiculous."

"Maybe so, but I'm still better than some stupid moving cat. Why should I care what a moving cat sees?"

"The principle of relativity is useful for things other than converting between frames. You can use it to make a complicated problem that you don't know how to solve look like a simple one you do know how to solve."

"Yeah? How?"

"Well . . ."

RELATIVITY IN RIVER CITY: USING THE PRINCIPLE OF RELATIVITY TO UNDERSTAND COLLISIONS

The principle of relativity may seem like an abstract exercise, but in fact it's an essential tool for physics. If you know how to solve a simple problem in a stationary frame, you can often solve more complicated problems by moving to a frame of reference in which the complicated problem resembles the simple problem you know how to solve. You then use your knowledge of the simple problem to determine what happens in the moving frame, and you use the Galilean recipe for converting back to the stationary frame you started in to get the answer you need.

The easiest way to see this in action is in the context of collisions. It's difficult to come up with a good canine example for this, but in a direct head-on collision between a moving object and a stationary object of equal mass—

"The moving object stops, and the stationary one moves off with the same speed and direction as the original moving object."

"Well, yeah. How do you know that?"

"It happens all the time in pool."

"Wait, pool? You play pool?"

"Of course not, silly, the tables are way too high. Plus it's hard to work the cue with my mouth."

"How do you know anything about pool, then?"

"Oh, it's on cable a lot in the middle of the afternoon. It's very relaxing to watch on days when I don't feel like reading your physics books."

". . ."

"What? It's boring when you're not there!"

As any billiards-loving dog knows, when a moving object, such as a cue ball, hits a stationary object of equal mass, such as the eight ball, after the collision the eight ball will move off in the direction that the cue ball was headed, while the cue ball will come to a stop at the spot where it made contact with the eight ball. This is a simple example of a one-dimensional collision—the actual collision takes place in three dimensions, but as long as the cue ball hits the eight ball squarely, without spin, we can treat this like a situation where the balls only move in one dimension.

A more complicated situation also involving a one-dimensional collision between identical objects is the case where both balls are moving. This is still one-dimensional, provided the collision is head-on, but it's not as obvious what will happen. We can use the principle of relativity to understand this, though, by making the complicated case of two moving balls look like the simple case of one moving ball hitting one stationary ball.

"Unless, of course, your extensive TV watching has given you the answer to this one as well."

"No, you don't see that very often in pool. I have no idea what to expect."

"I'm going to start hiding the TV remote when I go to work, you know."

"Look, I can watch TV, or I can chew on the furniture. Your choice, dude."

Let's imagine the collision between two moving billiard balls as seen by a dog sitting at rest in the pool hall and a cat in motion. To be concrete, let's imagine the cue ball moving east at 10 m/s, while the eight ball is moving west at 5 m/s. Nero the cat sees that Emmy is engrossed in watching pool and takes this opportunity to slink across the room for some nefarious purpose, moving west at the same speed as the eight ball. In Emmy's frame of reference, the two moving balls come together, collide, and then move apart. It's not obvious, however, what the final speeds of the two balls should be—we can determine it using the physics of momentum and energy, but it takes a good deal of math.

We can simplify the problem tremendously, though, by asking what the collision looks like according to Nero. As he moves west at 5 m/s, the eight ball appears to be standing still, while the cue ball is moving not at its Emmy-frame speed of 10 m/s but at 15 m/s (the 10 m/s speed of the ball plus the 5 m/s speed of the cat moving in the opposite direction). In Nero's frame of reference, the collision looks exactly like the simple collision we understand from watching pool: the moving cue ball stops, while the stationary eight ball moves off at the 15 m/s speed of the cue ball.

The principle of relativity tells us that the laws of physics will work the same way in both Nero's moving frame and Emmy's stationary frame, as long as we convert from one to the other properly. We can use the simple result in Nero's frame, then, to predict what Emmy sees when both balls are moving. Nero sees the cue ball as stationary after the collision, which means that, in Emmy's frame of reference, it is moving west at the same speed as Nero, 5 m/s. Nero sees the eight ball moving east at 15 m/s, which means that in Emmy's frame of reference, it's moving east at 10 m/s (the speed Nero sees, minus his speed according to Emmy). In other words, when the collision takes place, the two balls exchange velocities—the eight ball moves off with the same speed and in the same direction as the incoming cue ball, and vice versa.

"That's pretty cool. Does that work for all collisions?"

"It's only true for head-on collisions between objects of the same mass. And even then, it only works if the collision is elastic—that is, if the two objects collide and separate without getting dented or losing energy in other ways."

"Oh. That's not that many collisions."

"No, but then this is just an illustration of the basic idea. You can use the same trick of changing frames to understand other types of collisions, though. For example, what happens if you toss a light racquetball off a stationary basketball?"

"Ummmm . . . the racquetball bounces back the way it came, and the basketball barely moves."

"Right. Now, what happens when the basketball is moving toward the racquetball at the same speed?"

"I have no idea."

"You can understand what's going to happen by looking at the situation from the point of view of somebody moving with the basketball. And you can demonstrate it by taking a racquetball, putting it on top of a basketball, and dropping both of them on the ground. Just after the basketball hits the ground, it moves up and collides with the racquetball coming down."

"So, in the basketball's frame, it's stationary, and the racquetball bounces off it. In the stationary frame, then, the racquetball's final speed is its initial speed plus the speed of the basketball. So it comes out moving faster than it went in!"

"Right. If you drop them just right, the racquetball shoots way up into the air. It's a good physics party trick."

"I bet. Especially since you get to chase the racquetball afterwards. Chasing bouncy balls is fun!"

The mathematical description of motion in different frames may seem complicated when you first encounter it, but if you back up and look at the big picture, it's just common sense codified into simple mathematical recipes. To convert the position measured by a stationary dog to that measured by a moving cat, you subtract the cat's velocity multiplied by the time from the dog's measurements along the direction of motion, leaving measurements in the other two directions unchanged. To convert the velocity measured by the dog to the velocity measured by the cat, you subtract the cat's velocity from the dog's measurements along the direction of the cat's motion, leaving the other two directions unchanged.

When Einstein began thinking about the physics of moving objects in the late 1800s, this combination of Newton's laws and Galileo's principle of relativity was still the state of the art as far as moving objects were concerned. The laws of physics had expanded to include a great many new phenomena unknown to Newton, though, and these proved more difficult to reconcile with the Galilean idea of relativity. As we'll see in the next chapter, the new physics of electromagnetism conflicted with the commonsense Galilean view of relative motion, a conflict that common sense would ultimately lose.

"Wait a minute—so all that stuff with adding velocities was a lie? Why did we spend all this time writing it down?"

"It's not a lie, just an imperfect approximation. Galilean relativity works very well for essentially all everyday situations, but it's not the complete picture. And it breaks down when you start to think about really fast motion, and especially when you start to think about electromagnetism."

"How can it break down, though? I mean, it's all so obvious, it has to be right."

"It seems really obvious, but there's a hidden assumption being made in the Galilean case that turns out to be incorrect. Einstein's key contribution was to point this out."

"What's the assumption?"

"That's in the next chapter—you'll have to wait for it."

"Humph. I don't like waiting. I want to get to the part where I convert squirrels into energy!"

"This is important background information. You need to understand what the historical notion of relativity was before you can understand where it went wrong and how Einstein's theory fixes it. Anyway, your other plan was apparently to watch TV, so it's not like you were going to do anything better with the time."

"I could've been chasing squirrels. Or bunnies. Or even my tail."

"I don't think that counts as a better use of your time."

"Sure it does. Wheeee!"

Chapter 2

FAKE PROOFS AND FAILED EXPERIMENTS: HISTORICAL ORIGINS OF RELATIVITY

I'M WATCHING *PARDON THE INTERRUPTION* AFTER WORK, and they're talking about the Belmont Stakes. They show a clip of horses running, and Emmy pipes up, "I like horses!" She does this whenever she feels I'm not paying her enough attention.

"Horses are okay," I say.

"Okay? Horses are really neat!" She thumps her tail on the floor to emphasize the point.

"I guess." An evil idea comes to me. "Say, did you know that horses have an infinite number of legs?"

"What?"

"Yeah," I say, pausing the DVR. "Horses have an infinite number of legs, and I can prove it with logic."

"How?"

"Well, we know that horses have an even number of legs, right?"

"Well, yeah."

"And we also know that horses have both forelegs and back legs, right?"

"Sure."

"Now, four legs plus two back legs is six legs, which is an odd number of legs for a horse."

"I guess, but . . ."

"We already said that horses have an even number of legs. And the only number that is both even and odd is infinity. Therefore, all horses have an infinite number of legs."

She stares at me with huge eyes. "Whoa, I think you just blew my mind."

"Pretty good, huh?"

"Dude, I had no idea . . ." She continues to look amazed. "But why do they look like they only have four legs on TV?"

"No, no—it's a joke. They really do only have four legs."

"But . . . you used logic. . . . What about the back legs?"

"It's not really logic. It's a kind of Groucho Marx fast-talking fake logic. It puns off 'forelegs,' meaning the two front legs, and acting as if it means 'four legs,' giving the number of legs."

"Oh." She still looks confused. "Why would you do that, though?"

"It's an old math joke. There's a whole bunch of joke math proofs out there if you look for them. There's a bunch of different ways to 'prove' that 1 = 2, as well." I do the wiggly fingers for the quotation marks, just to make sure she gets the point. Being a dog, she sometimes misses those little details. It's very trying.

"Why would anybody want to do that?"

"Well, for one thing, it's funny." She cocks her head sideways. "Okay, fine, it's funny if you have the right sort of personality. It's also a useful reminder about the importance of checking all your assumptions."

"How's that?"

"Well, the proofs that 1 = 2 rely on slipping in some illegitimate operation in a way that looks plausible. You always end up dividing by zero, or something like that, but if you're not paying close attention, you can miss the incorrect step."

"I guess that makes sense. But how many reminders to be careful about math do humans need, anyway?"

"More than you might think. Whole fields of math and physics have been created by people realizing that they were assuming something that wasn't true."

"Like what?"

"Well, relativity, for example. One of the key realizations in Einstein's special theory of relativity is that time is different for different observers. Prior to 1905, people kept running into contradictions when they thought about light emitted by moving objects. Common sense says that light from moving objects should move at a different speed than light from stationary ones, but **Maxwell's equations** of electromagnetism give you a single, constant speed of light, no matter what speed the source or the observer has."

"That doesn't seem right . . ."

"It's weird but true. It doesn't seem right because you're making the same hidden assumption as the nineteenth-century physicists who were struggling with the problem: you're assuming that there's a single, universal time that all observers can agree upon."

"There isn't?" She looks confused.

"It looks that way, but when you think really carefully about how to measure time, it turns out that there is no such thing as an absolute time. Special relativity flows from the realization that different observers will disagree about the timing of events. Once you have that idea, you can work out the rest of it relatively easily."

"Pardon the pun . . ."

"Yeah, pardon the pun. Anyway, the point is, joke proofs are useful. In addition to being funny."

"I still don't see the humor, but whatever."

"Right, whatever." I sit back again, and hit play. Mike and Tony resume talking about racing.

"Anyway, as I was saying, horses are neat!"

"Even if they don't have an infinite number of legs."

"Yeah," she thumps her tail enthusiastically. "Their crap smells fantastic!"

I don't know what to say to that, so I just don't say anything.

Every profession has its share of inside jokes that only make sense to members of that profession, and physics and math are no exception.* Some of the jokes in mathematics serve an educational purpose, though—they're as much puzzles as jokes, reminders of the importance of thinking clearly and completely about a problem and the assumptions made in trying to understand it.

The joke proof that opens this chapter relies on fast-talking wordplay and exploits the natural tendency of dogs and humans to believe that one sentence will follow logically from the preceding sentences. Real fake proofs are more formal but ultimately rely on a similar mistake—making an incorrect assumption so basic you don't even notice it.** This sort of error in thinking can trip up even professional physicists and mathematicians, leading to apparent paradoxes. Only when the incorrect assumption is identified and corrected do we understand what's really going on, at which point the paradoxes turn out not to be paradoxical at all. This theme of hidden assumptions leading to apparent paradoxes runs through all of physics, but it's particularly pronounced in relativity, as we'll see many times in coming chapters.

In many ways, the problematic situation of physics at the end of the nineteenth century is similar to that of a student encountering one of these joke proofs in mathematics. By the 1800s, the laws of motion and the mathematical machinery for describing motion laid out by Galileo and Newton were well enough established to seem like the codification

* The ultimate only-funny-to-a-mathematician joke: "Q: What's purple and commutes? A: An Abelian grape."

** For example, there's a "proof" that 1 = 2 at www.math.toronto.edu/mathnet/falseProofs/first1eq2.html.

of common sense. New physics discovered in the middle of the century, though, seemed to set up a paradox, particularly when new experiments began to call into question some cherished assumptions about the nature of motion.

In this chapter, we discuss how the new physics of electromagnetism creates an apparent paradox when using the Galilean recipe for converting from one frame of reference to another. We also talk about the greatest failed experiment in the history of physics and how it set the stage for Einstein and others to correct the hidden assumption at the heart of classical physics and to reconcile the apparent paradox—at the cost of a complete rethinking of the nature of space and time.

NEW PHYSICS FOR THE NINETEENTH CENTURY: ELECTRICITY AND MAGNETISM

In Galileo's day, the study of physics was more or less restricted to looking at the motion of objects, but by the end of the nineteenth century, a whole new branch of physics had been established: the study of electricity and magnetism (affectionately known as "E&M" to physicists), codified in Maxwell's equations. E&M is concerned with the behavior of charged particles and magnets, and while these may seem like very different phenomena, in 1861 the full mathematical description of both electricity and magnetism was published by the Scottish physicist James Clerk Maxwell. This theory shows that electricity and magnetism are, in fact, different aspects of the same fundamental interaction and that light can be explained as an **electromagnetic wave.**

Maxwell's theory of electromagnetism consists of four short equations.* If you know the right sort of nerds, you may have seen them on a T-shirt or coffee mug in the following form:

---

* Interestingly, the four equations now associated with Maxwell's name were mostly developed by other people. The first two are forms of Gauss's law applied to electric and magnetic fields, respectively. The third equation is Faraday's law, and the fourth is Ampere's law plus the one small but crucial term that Maxwell himself contributed. He was the first, however, to recognize that these four equations represented a complete description of electricity and magnetism and to present them all together, so he gets credited with the whole lot. Timing is everything.

"God said:

$$\nabla \cdot \vec{E} = \frac{\rho}{\epsilon_0}$$

$$\nabla \cdot \vec{B} = 0$$

$$\nabla \times \vec{E} = -\frac{d\vec{B}}{dt}$$

$$\nabla \times \vec{B} = \mu_0 J - \mu_0 \epsilon_0 \frac{d\vec{B}}{dt}$$

And there was light."

"Umm, dude, the Genesis reference is cute and all, but I'm not getting a lot out of these equations."

"You don't need to understand them in detail. They're just decorative."

"I can think of a lot of things that would be more decorative. Pictures of me, say. Or pictures of big fat bunnies that I could chase. Or pictures of steak. Or actual steak."

"The logistical issues of putting actual steak in a book are a little beyond me. But as equations go, you have to admit that these are kind of pretty."

"Has anyone ever told you you're an incredible nerd?"

"I'm a physicist. That's pretty much a given."

"True enough."

While Maxwell's equations may look intimidating in mathematical notation, they're actually very simple to translate into English. They're a set of rules for determining the behavior of electric fields (represented by E) and magnetic fields (represented by B). If you know the electric and magnetic fields, you can use them to predict everything about the motion of charged particles like **electrons** or **protons.** The four equations translate into four rules:

1. The strength of an electric field depends on the amount of charge in the vicinity.
2. A magnet will always have both north and south poles.
3. An electric field is created by a changing magnetic field.

4. A magnetic field is created by either a current or a changing electric field.

These four rules tell you absolutely everything you need to know about electric and magnetic fields, which in turn tells you everything you need to know about the behavior of charged particles and magnets. You can use Maxwell's equations to understand how static electricity makes balloons stick to the ceiling after they're rubbed on a dog's fur, or how electric current flows to power all the appliances humans have for storing and preparing food, and even how to use spinning turbines attached to magnets to generate the electricity needed to maintain comfortable homes for humans and dogs.

You can also use Maxwell's equations to understand light. This might not seem immediately obvious, but it follows rather simply from the third and fourth rules. In the absence of any current flowing, these state,

3. An electric field is created by a changing magnetic field.
4. A magnetic field is created by a changing electric field.

Going back and forth between these two lets you create a mutually sustaining electromagnetic wave: A changing magnetic field creates an electric field. That (changing) electric field creates a magnetic field. That magnetic field, in turn, creates an electric field, which creates a magnetic field, and so on. The end result is an oscillating electric field coupled with an oscillating magnetic field, which move along together through space in a direction perpendicular to both the electric and magnetic fields. An electric field oscillating vertically, sometimes pointing up and sometimes down, paired with an east-west magnetic field will move either north or south.

In 1887 the German physicist Heinrich Hertz demonstrated the existence of oscillating electromagnetic waves traveling through space by mapping out the waves generated by an electric spark.* The unit of frequency for an oscillating system is named in his honor: one hertz (Hz) is

* In the course of these experiments, Hertz also made the first observation of the photoelectric effect, which Albert Einstein would use in 1905 to show that light behaves like a particle as well as a wave.

one oscillation per second. While Hertz himself did not fully appreciate the meaning of his accomplishment,* these waves were soon identified as light, providing a nearly complete description of classical optics as well.

Maxwell's equations are a great triumph of nineteenth-century physics, providing a complete mathematical theory of charged particles and light. Together with the laws of thermodynamics, which were developed at the same time, and Newton's well-established laws of motion and gravitation, they let many late-nineteenth-century physicists feel as if they were on the verge of a complete description of the universe. In many books, you can find some famous physicist of that time quoted as saying, "There is nothing new to be discovered in physics now. All that remains is more and more precise measurement."** In fact, physics was on the verge of two major crises as it entered the twentieth century, each crisis requiring a revolutionary new theory to resolve the problem. One crisis, having to do with the nature of matter, led to the development of quantum mechanics, which is a subject for a different book.† The other crisis, having to do with the nature of space and time, led to the theory of relativity, which is the story we'll be telling here.

LIGHT-CARRYING WAVY STUFF: THE LUMINIFEROUS AETHER AND THE SPEED OF LIGHT

Einstein's first paper laying out the special theory of relativity is titled "On the Electrodynamics of Moving Bodies,"†† which may seem surprising given that we started our discussion by talking about moving cats and

---

* Asked about possible uses for his discovery, Hertz said, "It's of no use whatsoever. . . . This is just an experiment that proves Maestro Maxwell was right—we just have these mysterious electromagnetic waves that we cannot see with the naked eye. But they are there." This shows that physicists are not always astute judges of potential technological applications: within seven years of Hertz's experiments, Guglielmo Marconi had demonstrated the use of these waves for carrying radio signals.

** The quote is most often attributed to the great British scientist Lord Kelvin, but there doesn't seem to be a concrete record of him (or anyone else) ever saying this.

† *How to Teach Physics to Your Dog* (New York: Scribner, 2009).

†† Actually, it's "Zur Elektrodynamik bewegter Körper," because it was published in the German journal *Annalen der Physik* 17, no. 891 (1905). Physics didn't become dominated by English-language journals until after World War II.

dogs, but the theory of relativity is intimately connected with electromagnetism. In fact, Maxwell's equations contain the seeds of the crisis to which relativity is the solution.

As described above, the third and fourth Maxwell equations let us create electromagnetic waves consisting of oscillating electric and magnetic fields that support each other as they travel through space. The full mathematical treatment of this gives the speed at which these waves travel, traditionally given as the symbol c, which is a very simple mathematical relation:

$$c = \frac{1}{\sqrt{\mu_0 \varepsilon_0}}$$

The speed of light predicted by Maxwell's equations is one over the square root of the product of two constants of nature, μ_0 and ε_0.*

What's most important about this equation is what's *not* found in it—namely, any reference to the speed of the source of the waves or the observer. Maxwell's equations predict the existence of electromagnetic waves, now identified with light, but they don't provide any way of including the effects of motion. This seems to put Maxwell's equations in direct conflict with the principle of relativity.

As we saw in the previous chapter, two observers who are moving with respect to one another—a dog in the window and a cat in a passing car—will disagree about the speed of a moving object. If Emmy the dog sees a child on a bicycle moving west at 6 m/s and Nero the cat moving east at 10 m/s, Nero will say that the child is moving at 16 m/s—the speed seen by Emmy, plus his speed relative to Emmy.

Simple logic suggests that this Galilean recipe for converting from one frame of reference to another ought to carry over to the case of light waves. That is, if the bicycling child turns on her headlight, she should see the light leaving the headlight at the speed of light, Emmy should see the light leaving the lamp moving at the speed of light plus 6 m/s (the speed of the bike relative to Emmy), and Nero should see the light coming toward him at the speed of light plus 16 m/s (the speed of the bike relative

* These are, respectively, the "permeability of vacuum" and the "permittivity of vacuum," but the names and their exact values don't really matter for our purposes.

to Nero). And yet, Maxwell's equations don't provide for a variable speed of light: according to the equations, light always and everywhere moves at exactly the same speed, no matter how fast the source of the light is moving or how fast the observer seeing the light is moving.

"Wait, doesn't it depend on what's waving?"

"What do you mean?"

"You know, what carries the waves. Like, the waves in the pond in the backyard are carried by the water in the pond, and the sound waves when I bark are carried by the air in the house. Light waves must be carried by some sort of light-carrying wavy stuff."

"The **luminiferous aether**."

"Ooh! That's a great name. What's it mean?"

"'Light-carrying wavy stuff,' more or less. In Latin."

"Latin sounds so much cooler than English. You should teach me Latin."

"I'll get right on that. Anyway, you've hit on one of the great historical attempts to explain the propagation of light. 'Luminiferous aether' is the name of the mysterious substance that was imagined to carry light waves even before Maxwell's equations came along, and it provides a possible way to weasel out of the problem of the single speed of light."

"How is that?"

"Well, if you imagine that light waves are propagating through some aether, that aether must fill the entire universe, since we see light waves coming to us from very distant stars and galaxies."

"That's a lot of aether, isn't it?"

"At the time, nobody knew how incredibly mind-bogglingly huge the universe really is, but even so, you're right that it's a lot of space to fill with stuff. The aether itself would need to be extremely light, though, so it might not amount to all that much mass. Anyway, the idea is that the light waves always propagate through this aether that is everywhere and fixed in space. If that's the case, then the single speed predicted by Maxwell's equations would be the speed of light waves with respect to the aether."

"So, no matter how fast you were moving when you flipped on the lights, the light would travel through the aether at the same speed."

"Right. If you use that model, then you could say that the speed of light relative to the aether was fixed, and the speed of light according to any observer would depend on the speed of that observer relative to the aether—which you could figure out using the Galilean rules we talked about last chapter."

"That's great! I fixed the problem! I'm a genius dog."

"Not so fast. The aether model was popular, but it had a lot of problems."

"Like what?"

"Well, for one thing, the aether would need to be absurdly light—the British physicist Lord Kelvin estimated the density of the aether and came up with a figure of 10^{-21} kilograms (kg) per cubic centimeter (cm^3). That's the mass of only about a million hydrogen atoms."

"A million's kind of big, though, right?"

"If you're talking about dogs or people, not atoms. That's around a hundred trillion times less dense than air."

"Oh."

"It would also need to be fantastically stiff—way more rigid than any known substance—but at the same time allow ordinary matter to pass through it almost undisturbed."

"That's . . . really weird. Kind of hard to believe, really."

"Exactly. More importantly, people have done lots of experiments looking for it, one of which we'll talk about next, and never found any hint of it—which is why nobody really believes in it as a theory of physics anymore."

"OK, so it doesn't solve the problem. It's still good for one thing, though."

"What's that?"

"Luminiferous Aether would be a great name for a band."

THE WORLD'S GREATEST FAILED EXPERIMENT: THE MICHELSON-MORLEY EXPERIMENT

Any time two theories of science come into conflict, the resolution is always achieved through experiments. The business of science, particularly physics, is to provide a mathematical description of reality, and the ultimate test of any theory is whether its predictions are borne out in reality. In this case, the two theories make very distinct predictions: Maxwell's equations say that light has a single speed independent of the motion of the observer, while aether theories predict that the observed speed depends on the observer's motion relative to the aether. This question can be settled experimentally by measuring the speed of light in different conditions.

Of course, measuring the speed of light is a challenge even today, as the speed of light is incredibly fast—just under 300 million meters per second.* Measuring changes in that speed due to the motion of macroscopic objects requires both a really fast-moving test object and an extremely sensitive measurement technique.

The most famous experiment to look for the effects of motion on the speed of light is known to physicists as the **Michelson-Morley experiment**; it was carried out by American physicist Albert Michelson** with the assistance of Edward Morley, starting in the 1880s. It is widely regarded as one of the most important and influential experiments in the history of physics, despite the fact that it failed in its original purpose.

For their moving object, Michelson and Morley chose the fastest-moving object available to them, which also happens to be the fastest-moving object available today: the motion of Earth in its orbit around the sun. Earth orbits the sun once per year at a distance of around 150 million kilometers (km), which works out to a speed in its orbit of around 30,000 m/s.† The

* In a happy coincidence of units, the speed of light is very nearly one foot per nanosecond, one of the few cases in science where the imperial units commonly used in the United States are more convenient than metric units.

** Michelson had a fascinating life: born in modern-day Poland, he moved to America with his parents as a very small child and grew up in gold rush towns in California. He managed to get an appointment to the Naval Academy in Annapolis by traveling across the country to personally request one from President Ulysses S. Grant. Grant had already used up his allotment of appointments but was sufficiently impressed with Michelson that he granted him admission anyway, an "illegal act" that Michelson later chuckled over. He went on to a distinguished career in physics and became the first American awarded a Nobel Prize in science, winning the 1907 physics prize for his precision measurements of light.

† Earth's orbit is slightly elliptical, so the speed changes slightly as it moves through the year. The difference between the closest point to the sun and the farthest point from the sun is only about 5 million kilometers, though, so it doesn't make a great deal of difference.

fastest man-made object in history is the Voyager I space probe, currently leaving the solar system at about 17,000 m/s, just over half the speed of the orbiting Earth.

As fast as it is, though, the speed of the Earth is still only 1/10,000th that of light. In order to measure changes in the speed of light at that level, Michelson needed to invent a new measurement technique that makes use of the wave nature of light to attain incredible sensitivity.

The center of a **Michelson interferometer** is a beam splitter, a glass plate that lets half of the light falling on it pass through, while the other half is reflected. The beam splitter is arranged so that it splits a beam of light into two pieces that follow perpendicular paths (if the light is initially headed east, half of it continues to the east, while the other half is reflected to the north). Each of these paths ends at a mirror that reflects the light back on itself, returning to the beam splitter. Half of each beam is sent toward a detector by the beam splitter (half of the light from the north-south arm is transmitted, while half of the light from the east-west arm is reflected to the south), where the intensity of the light is recorded.*

You might expect that the detector would always see half of the input intensity (it receives two beams that are each one-quarter the original intensity), but in fact, the intensity at the detector can range from zero to the full intensity of the light source, thanks to a wave phenomenon known as interference. When we add together the light waves, what we get depends on whether the peaks of one wave fall in the same places as the peaks of the other. If the peaks line up, as shown in Figure 2.1, the resulting wave is twice as large as either of the original waves (this is called *constructive interference*). If the peaks of one wave fall in the valleys of the other, though, the waves cancel each other out, leaving no light at all (*destructive interference*).

This interference effect makes the Michelson interferometer an incredibly sensitive method for measuring small differences between the two arms of the interferometer. If you move one mirror farther from the beam splitter

---

* In the Michelson-Morley experiment, the detector was a person looking through a telescope. Nowadays, we monitor the intensity of the light with electronic detectors, which are cheaper and don't need as many breaks.

Figure 2.1. A Michelson interferometer, showing constructive interference of the light waves. Light from the source hits a beam splitter, and half of it travels up to the top mirror, while the other half continues straight through to the mirror on the right. The mirrors direct the beams back to the beam splitter, where half of each beam is sent downward toward the detector. When the two waves follow paths of the same length, as they do here, the waves from the two paths interfere constructively, so the peaks of one fall on top of the peaks of the other, making a bright spot at the detector.

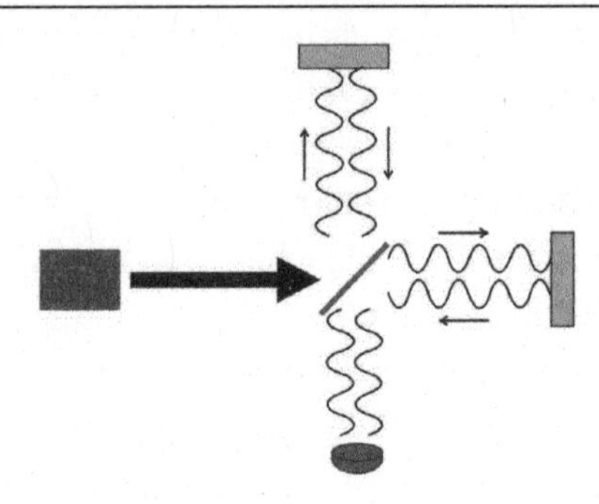

by just one-quarter of the wavelength of the light,* as shown in Figure 2.2, the light on that path travels an extra half wavelength. During this extra travel time, the wave completes another half oscillation, so when the two waves are combined, the peaks of one fall in the valleys of the other. Moving another one-quarter of the wavelength will return to constructive interference, giving a bright spot at the detector: the wave on the longer path completes one full extra oscillation, so the peaks overlap again, and so on. If you continue to move one mirror back, you will see a series of bright and dark spots, referred to as *interference fringes*. Counting the number of these fringes that go by when moving a mirror from one place to another lets you measure changes in the mirror's position to within a few nanometers** (nm).

The same interference effect that makes the interferometer a sensitive detector of changes in the length of one of the two arms also allows it to measure changes in the speed of light. Light passing through a transparent substance like air or glass moves slower than light traveling through empty

* If you're using visible light, this works out to a bit more than 100 nm, or about 1/1,000th the thickness of a human hair.

** One nanometer equals 0.000000001m.

Figure 2.2. The same Michelson interferometer, with one mirror moved back by one-quarter of a wavelength. The extra distance traveled by the light hitting that mirror puts the peaks of that wave in the valleys of the wave from the other path, causing destructive interference and a dark spot at the detector.

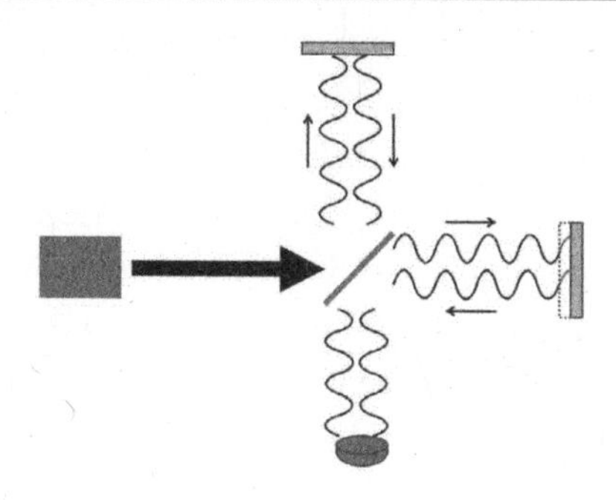

space. If you remove all the air from one arm of a Michelson interferometer, leaving the other arm alone, the light at the detector goes from bright to dark and back many times as the air is pumped away. The number of these interference fringes tells you the difference between the speed of light in air and the speed of light in a vacuum, which is only about 0.03 percent of the speed (90,000 m/s) but is readily observable using an interferometer.

The Michelson-Morley experiment used an interferometer to look for motion of the Earth through the luminiferous aether. If the single speed of light predicted by Maxwell's equations is the speed of the light relative to the aether, then as the Earth moves through the aether, light traveling along the direction of motion must travel at a different speed relative to the Earth than light moving at right angles to the motion. Of course, there's no way to stop the Earth's motion to do the comparison, but Michelson and Morley found an ingenious way around that problem: they set their interferometer up with one arm pointed in the direction of the Earth's motion, then rotated it by 90 degrees. After the rotation, the arm that had been along the motion was now perpendicular to it, while the arm that had been perpendicular to the motion was now aligned along the motion. If light traveling out and back along the direction of Earth's motion moved at a different speed than light traveling at right angles to the motion, rotating the interferometer should exchange the roles of the two arms. Rotating the arms should produce a shift in the amount of light at the detector as the speed changes from one value to the other, exactly as happens when air is pumped out of one arm.

Michelson and Morley built a huge interferometer, with arms 11m in length, and mounted it on a large block of granite to minimize any chance of vibrations moving the mirrors. They floated their two-ton granite block on a vat of mercury to ensure smooth rotation of the apparatus* and repeated the experiment at different times of day and different points during the year. They measured no significant change in the speed of light—a very conservative interpretation of their measurements said that the Earth's velocity relative to the aether had to be less than one-sixth of the orbital speed of the Earth.

"But, wait, doesn't that just mean that the aether is moving with the Earth?"

"That's one possible way to explain the negative result—that the Earth drags the aether in its vicinity along with it, so that the relative speed of the Earth and the aether is much smaller than it would be otherwise. This runs into problems with other measurements, though, looking at the positions of stars at different times of the year."

"What's that got to do with anything?"

"The motion of the Earth relative to the light coming from a distant star causes a small shift in the apparent position of the star that depends on the direction of motion. This **stellar aberration** means that stars sort of wobble around during the course of the year. This wobble had been observed back in the 1700s and was well established by the time Michelson and Morley did their experiments."

"How does Earth's motion make stars move?"

"Well, it's sort of like walking in the rain with an umbrella. Even if the rain is falling straight down, you need to tip the umbrella forward a little bit because your motion relative to the rain makes it look like the rain

* You'd never be able to get away with this today, but OSHA didn't exist in the 1880s.

comes in with a little horizontal velocity. In the same way, the motion of the Earth relative to the light requires you to tilt the telescope at a slight angle, which makes the star look like it's in a different position. If the Earth was really dragging the aether along with it enough to explain the Michelson-Morley result, you would see a smaller aberration than we do."

"Okay, I guess. There's one problem with your explanation, though."

"What's that?"

"When we walk in the rain, you're the one holding the umbrella, not me. No matter how you tilt it, I still get soaked."

"Oh. Well, sorry about that."

The Michelson-Morley experiment is arguably the greatest failed experiment in the history of physics. Morley himself didn't believe their initial results and spent the next couple of decades putting together better and better tests, but numerous repetitions of the experiment over the last 125 years have confirmed the initial negative result. The speed of light is independent of the motion of the Earth to within a small fraction of a meter per second.

The Michelson-Morley experiment was a significant step in forcing physicists to accept that Maxwell's equations accurately describe reality: light moves through empty space with a single speed, no matter what the observer is doing. This requires a dramatic modification of the way that we look at motion in different frames of reference, which seems to run counter to common sense in the form of the Galilean rules we discussed earlier.

The mistaken assumption that led Michelson and Morley to think they could measure Earth's speed relative to the aether is subtle and difficult to spot, like the incorrect step in a good fake mathematical proof, and it hinges on the nature of time. The interferometer is sensitive to motion through the aether because a different speed of light along the direction of motion changes the time required for light to travel out and back along that arm of the interferometer. This implicitly assumes, however, that this time is the same for all observers—that an experimenter moving with the interferometer and a hypothetical dog at rest with respect to the aether watching the Earth zooming past will measure the same round-trip time.

That may seem too obvious to count as an assumption, let alone an incorrect one, but it's the crucial flaw. If the observer with the apparatus sees time passing at a different rate than the dog in the aether, then the negative result of the Michelson-Morley experiment can be explained as a logical consequence of this behavior. Correcting this mistaken assumption is the key step that underlies the entire theory of relativity.

The Michelson-Morley result was also the first step in the process of removing the luminiferous aether from modern physics. The properties required for such a substance were already known to be problematic, and the failure to measure any motion relative to the aether helped convince physicists to abandon the idea altogether. Modern physics holds that electric and magnetic fields do not require any medium to support them but exist on their own in empty space. The various Michelson-Morley tests over the years haven't found any motion relative to the aether for the very simple reason that there is no aether to move relative to.

The mathematical tools needed to understand the Michelson-Morley result were developed not long after the experiments by Irish physicist George FitzGerald and Dutch physicist Hendrik Lorentz. The necessary techniques produce some very strange results—in addition to disagreement about timing, the interferometer must shrink along the direction of motion—so they weren't generally accepted until Einstein published his first relativity papers in 1905, almost twenty years after the Michelson-Morley result. Einstein succeeded where others had failed by showing that a careful treatment of time and motion make these effects inevitable. The strange phenomena are not a weird and arbitrary mathematical fiction but the unavoidable result of the fundamental structure of our universe.

"Wait—Einstein didn't come up with relativity on his own?"

"No. Other people worked out all of the mathematical apparatus before him. Hendrik Lorentz worked out the equations that physicists use to correctly determine what moving observers see, now called the Lorentz transformation equations in his honor. FitzGerald predicted the phenomenon of **length contraction,** which we'll talk about in Chapter 4, and the French physicist Henri Poincaré almost put the whole thing together a little before Einstein."

"So, why is Einstein all famous, while I haven't heard of these other guys?"

"Because they all balked at the weirdness of the predictions, so none of them quite got it right. Poincaré was the closest, but even he stubbornly stuck with the aether idea of defining a preferred frame of reference for the motion of light and treating all the rest of the predictions as convenient fictions needed to make the math work out. Einstein was the first to put the whole package together, and more importantly, he made a convincing argument that this was how things *have* to be."

"Yeah? What's the argument?"

"That's what's in the next chapter."

"You're enjoying dragging this out, aren't you?"

Chapter 3

TIME SLOWS WHEN YOU'RE CHASING BUNNIES: RELATIVISTIC TIME DILATION

WE'RE JUST STARTING OUT ON A WALK, and no sooner do I open the backyard gate than Emmy takes off at a run, hitting the end of the leash and nearly pulling my arm out of its socket.

"Whoa, there," I say. "Take it easy."

"Come on!" she says. "We need to go fast! Let's go, let's go, *let's go!*"

"What's the hurry? It's a nice day—there's no rush."

"We need to go fast. If we go fast, I'll be younger than the human puppy." (My daughter is a toddler. It took several tries to explain the relationship, and Emmy now insists on thinking SteelyKid* is a puppy.)

"You know, I'm all in favor of exercise, but I think you're overstating the benefits."

"No, silly, it's not about exercise. It's physics. When I go fast, time slows down. It's simple relativity."

"The name is special relativity, actually. And this plan of yours is not going to work."

"You're not going to try to claim that moving clocks don't run slow, are you? Because they do."

"No, you're absolutely right. Moving clocks run slow." We stop at the corner to let a car pass.

"Right, because of triangles. Let's go!" Emmy starts pulling hard on the leash the instant the car is gone.

"Triangles?"

"You know. The light in a stationary clock bounces up and down, while the light in a moving clock goes in triangles. So it's slower."

"You're talking about the **light clock** thought experiment?"

"Yeah, it's one of my favorite *gedankenexperimenten*." She's part German Shepherd and loves showing off her language skills.

"You're garbling it a little, though. The idea is that you can derive the proper form for the **time dilation** effect that makes a moving clock appear to run slow by thinking about a clock based on light pulses bouncing between two mirrors. The light in a clock that's moving relative to you will follow a longer path than the light in an identical clock that's stationary. Since the speed of light is the same for all observers, that means that the time between ticks of the moving clock must be longer than for those of the stationary clock."

"Right. So if we go really fast on the walk, my clock goes all slow, and when we get back, I'll be younger than the puppy."

---

* This is the *nom d'Internet* we use when talking about her in public so that a Google search on her real name won't turn up embarrassing baby pictures when she's old enough to worry about such things.

"Well, no. For one thing, the effect only works on clocks that are moving relative to you. You would see a light clock at the house ticking slow, but SteelyKid would see *your* clock running slow."

"Yeah, but she's just a puppy and is easily fooled. My clock is the one that's really moving."

"No, she'd be right. As long as you're moving relative to her, any measurement she makes will show your clock running slow, and any measurement *you* do will show *her* clock running slow. There is no absolute frame of reference, just relative motion."

"Wait, we each see the other's clock running slow?"

"As long as you're moving at constant speed relative to one another, yes. And you're both right."

"So how do we decide which of us is younger?"

"Well, in a real experiment, the process of speeding up and slowing down makes your frame of reference distinguishable from hers, so if you take off from the house, accelerate up to very fast speed, then decelerate and return to the house, your clock will end up showing less elapsed time than hers, but—"

"I knew it! Let's go, so I can be younger than the puppy!"

"*But*, 'fast' in this context means 'at a speed comparable to the speed of light.' Even at our current—exceedingly brisk—walking pace, we're not going more than a few meters per second. Which means you would need to walk for a billion years to gain one second on SteelyKid."

Emmy stops dead. "A billion years?"

"A billion years."

"That's a long time." She thinks for a minute. "I'm good with that. I like long walks." She takes off again and just about pulls me off balance.

"I'm not good with that. I have things to do this evening. Anyway, why are you so fired up to age slower than SteelyKid?"

"Don't you want to? You know, skip by this boring phase and get to the part where she's old enough to use real words and talk about physics and help me catch bunnies and stuff?"

"Some of us think the current phase is pretty cute, you know."

"Yeah, well, humans are weird."

"Anyway, you're forgetting one important fact."

"What's that?"

"When she gets old enough to use real words, she'll be old enough to eat without spilling food—which means you won't get to eat all the stuff that she currently drops on the floor."

Emmy stops dead on the side of the road so quickly I almost trip over her. "Oooh. I didn't think of that. That's a good point."

"Thank you. Can we walk at a normal speed, now?"

"Sure. As long as the human puppy will still drop stuff for me to eat when we get back."

"I think that's a pretty safe bet." We resume walking at a more sedate pace.

"Good. I like puppy food!"

The crisis in physics precipitated by Maxwell's equations and the Michelson-Morley experiment forced physicists to reconsider some deeply held notions. The most fundamental of these notions was the idea of universal time. From Newton's day and even earlier, physicists believed in time as a universal constant. The entire universe was a clock of sorts, with each passing instant of time marked in the same way by all observers.

Confronted with experiments showing a constant speed of light, physicists were forced to go back to this most basic of assumptions and think about how it might be corrected. The solution was driven not only by scientific developments like the Michelson-Morley experiment but also by technological advances associated with railroads and transoceanic telegraph networks.

In this chapter, we look at the central principles of Einstein's 1905 special theory of relativity and what they mean for our understanding of time. We'll look at a simple thought experiment demonstrating one of the most surprising predictions of relativity—that moving clocks run slow. We'll also talk about some of the experiments that confirm this prediction to amazing precision.

A MATTER OF PRINCIPLES: THE SPECIAL THEORY OF RELATIVITY

Einstein published his special theory of relativity* in 1905, one of four revolutionary papers he produced that year. The core of the theory is surprisingly simple, consisting of just two principles:

> **Principle of relativity:** All of the laws of physics work the same way for any two observers moving with constant relative velocity.
>
> **Principle of the constancy of the speed of light:** All observers will measure the same speed of light c, regardless of their velocity relative to the source or to each other.

The first principle is just a generalization of the principle of relativity introduced by Galileo in Chapter 1. The wording is slightly different to emphasize the fact that the laws of physics include Maxwell's equations for electromagnetism as well as Newton's laws of motion. We also got rid of the words "stationary" and "moving," which imply a privileged frame of reference that is not moving: since the Michelson-Morley experiment shows that there is no aether, there is no such stationary frame. The motion of the observers relative to one another is all that matters.

The second principle is somewhat redundant: since the first principle includes Maxwell's equations, which predict a single speed for light, that necessarily means that all observers will measure the same speed of light. If they didn't, that would violate the principle of relativity. The notion of light having a constant speed for all observers was sufficiently revolutionary in 1905, though, that Einstein included it as a second principle for emphasis.

---

* It is "special" because it deals only with a special case, namely, the case of inertial frames of reference. The general theory of relativity, described in Chapters 9 and 10, adds the effects of gravity and acceleration and applies to essentially every imaginable situation.

Today, the theory of relativity is sufficiently well established that the speed of light in a vacuum is a defined quantity: 299,792,458 m/s exactly. In fact, the definition of a meter is now based not on any physical artifact but on the speed of light: it is the distance covered by light in 1/299,792,458th of a second.* Every time you measure the length of something, you are using relativity through the constancy of the speed of light.

From these two simple rules, Einstein developed an entirely new way of looking at space and time. More importantly, he showed that the strange consequences of these new rules follow very naturally from a consideration of the nature of space and time and how we measure them.

A THING THAT TICKS: CLOCKS AND THE NATURE OF TIME

A concise answer to the question, What is time? might be, Time is what we measure with a clock. This sounds tautological, but when it comes to thinking about relativity, this sort of operational definition is critical. Many of the essential elements of relativity are best understood by thinking about them in terms of the operation of clocks.

The obvious follow-on question, then, is, What is a clock? A concise answer to this question would be, A clock is a thing that ticks. That may seem a little strange in this age of digital clocks in cell phones, whose ticking isn't audible even to a dog with very good ears, but if you define "tick" broadly as "repeating some action in a regular way," it captures the essential operation of a clock. Any action that repeats in a regular way can be used as a clock: counting the number of repetitions that occur between two events gives you a way to measure the amount of time that has passed.

Over the past few thousand years, humans have used lots of different types of ticks to mark the passage of time with steadily increasing precision.

---

* This may seem circular, but it's really because our ability to measure time is vastly better than our ability to measure length. Atomic clocks provide an extremely precise definition of the second, and since we know that the speed of light is constant, we can use that ultraprecise timekeeping to give us an equally good length standard through the speed of light.

The first clocks used the motion of astronomical objects, tracking the hours of the day by the sun's position in the sky and the days of the year by noting where the sun rises and sets.* Later civilizations began to mark shorter intervals through the regular motion of earthbound objects, such as water clocks and hourglasses, which measure time by the draining and filling of containers. Galileo Galilei was the first to realize that a pendulum—such as a mass swinging from the end of a string**—provides an extremely regular oscillation that can be used to drive gears in a mechanical clock. The first pendulum clock was built by Christian Huygens in 1656, and over the next few centuries, scientists and engineers refined the idea of a pendulum clock, eventually building clocks that could keep accurate time to within a few hundredths of a second in a day.†

While that may seem excessive to dogs, who have little need for timekeeping beyond "Is it dinnertime yet?"†† the keeping of accurate time is of great importance for human civilization. Determining longitude at sea requires accurate timekeeping, so sailors demanded ever more precise clocks. And as railroads and telegraph lines began to carry messages faster and faster, time coordination—making sure widely separated clocks each had the correct time—became increasingly important and required ever better clocks.

In the early twentieth century, on discovering that quartz crystals vibrate at a very precise frequency, clock makers devised ways to use this oscillation to drive clocks. To this day, many humans wear quartz watches based on

---

* People promoting crazy theories about space aliens helping build ancient monuments will often assert that ancient humans couldn't possibly have made precise predictions about the motion of the sun without advanced technology. In reality, though, the technology needed is minimal—as Neil DeGrasse Tyson points out in *Death by Black Hole* (New York: W. W. Norton, 2007), all you need is a stick and a lot of patience.

** Legend has it that Galileo discovered the regularity of a pendulum's swing by watching chandeliers sway back and forth while he was bored in church. He timed their swing using his pulse as a reference and found that they completed the same number of swings in the same number of heartbeats whether they were swinging by a small amount or a large amount. This is perhaps the greatest discovery ever inspired by a dull sermon.

† There are 86,400 seconds in a day, so this is pretty remarkable.

†† The answer is yes. To a dog, it's always dinnertime.

this principle, which can be accurate to within a few seconds per year.* Even that is not enough for some humans, though. With the development of quantum mechanics, scientists realized that the frequency of light absorbed or emitted by atoms is set by the laws of physics. All atoms of a given element will absorb and emit exactly the same frequency of light, whose oscillations provide the tick for an **atomic clock.** The current standard for time measurements uses the light absorbed by a cesium atom moving between two particular states, which oscillates 9,192,631,770 times every second,** no matter where those atoms are or what they're doing. Using clever schemes to measure this frequency, humans have designed atomic clocks of truly astonishing precision—the best current clocks would gain or lose no more than one second in some 138 million years of continuous operation.

The history of clocks involves a huge range of different technologies for time measurement, but they all have one thing in common: a regular, repeated motion of some kind. Whether it is the sun moving from east to west across the sky, a pendulum swinging back and forth in a grandfather clock, or an oscillating electromagnetic field, all of these systems offer a regular tick to mark the passage of time.

A key element for relativity is that all of these various time measurements are absolutely equivalent to one another. No matter what tick you use for your clock, you will measure time passing at exactly the same rate.

"But that's not right!"

"What do you mean?"

"Well, your pendulum clock ticks once a second, but your atomic clock ticks 9 billion times a second. Nine billion isn't equal to one—even I know that, and I'm not good at math."

---

* There are roughly 31.6 million seconds in a year, making this even more remarkable than the pendulum clocks.

** As with the meter, this is a defined quantity: the second is 9,192,631,770 of these oscillations exactly.

"Right. But the relative rate at which they tick remains constant. That is, if the atomic clock ticks 9 billion times during the first tick of the pendulum clock, it will tick another 9 billion times during the second tick of the pendulum clock, and the third, and the fourth, and so on. The absolute number of ticks will be different, but always by the same factor of 9 billion."

"What good is a clock that ticks 9 billion times a second, anyway? I know I can't count that high, and I'm pretty sure you can't either."

"You don't actually count the individual oscillations. Atomic clocks use some fancy electronic techniques to reduce the output ticking to something more manageable that you can use to measure time or run a time display."

"Sounds complicated."

"It is, and that's just the start. The official atomic clock–based time for the world* is a consensus average of a bunch of different clocks run in different countries. They also add the occasional 'leap second' to keep the official time synched up with the rotation of the Earth. It's a fascinating subject, but a little off topic for us."

"OK, I guess. Let's talk about how chasing bunnies slows time."

TRIANGLES OF TIME: THE LIGHT CLOCK

In his 1905 paper introducing special relativity, Einstein demonstrates the changeable nature of space and time by considering the problem of time *coordination*, that is, making sure that two clocks in different locations read exactly the same time. Similar considerations led Henri Poincaré to develop a model of "local" time measured by moving observers, though he viewed this as a matter of convention and clung to the idea of a fixed aether. Peter Galison, in *Einstein's Clocks, Poincaré's Maps*, argues that this was no coincidence, that Einstein and Poincaré were both led to this approach because

---

* This refers to Coordinated Universal Time, which is stuck with the confusing abbreviation "UTC," thanks to a linguistic stalemate. At the international commission that established UTC, English-speaking countries wanted the abbreviation to be "CUT," while French-speaking countries wanted "TUC" (for "Temps Universel Coordonné"). They compromised on "UTC," which doesn't make sense in either language.

clock coordination was extremely important for their jobs. Poincaré, a well-established and respected scientist, philosopher, and statesman, was a member of several international commissions charged with establishing standards for time and longitude. From 1902 to 1909, Einstein worked as a low-level patent clerk* in Bern, Switzerland, which then as now was famous for its clocks. The Swiss patent office handled numerous patents for schemes dealing with coordination of widely separated clocks, so these issues would naturally have come to Einstein's attention during the course of his job.

While clock coordination was the original inspiration for relativity, we can more easily appreciate the fundamental relativistic effects by thinking about the operation of a "light clock," a thought experiment developed shortly after Einstein's work was first published. This clock consists of two parallel mirrors with a pulse of light bouncing back and forth between them. Every time the light pulse hits the bottom mirror, we record that as a tick of the clock.

The time between ticks of this clock is determined by the time required for light to make a round-trip, which is twice the distance between the mirrors divided by the speed of light. For a clock with a 1m separation between the mirrors, that works out to one tick every 6.7 nanoseconds (ns).** This is the time recorded by an observer at rest with respect to the clock, let's say Harley the poodle chewing on a purloined sock in the corner of the lab.

To understand relativity we need to ask, What is the time between ticks of an identical clock that is moving relative to our canine observer? To be

* Einstein was unable to secure a faculty position at a university and took a job with the patent office as a way to support his wife and young child while he completed his doctorate and looked for a permanent job. His inability to secure an academic position was not, as is sometimes claimed, due to any deficiency in his education. Academic jobs were no easier to obtain in 1905 than today, especially for a student who hadn't finished his dissertation, and Einstein also had to contend with the appalling level of anti-Semitism in Europe at that time.

** That's 0.0000000067 seconds, which seems awfully short but is sixty-one times longer than the time for an oscillation of the light used in an atomic clock and many millions of oscillations of visible light.

specific, let's think about a vertical light clock moving in a horizontal direction, for example, one carried by Desdemona the cat (Mo to her friends) cruising past Harley's clock at high speed. The recipe we use to determine the time between ticks of the moving clock is the same: the time is the total distance traveled divided by the speed of light. However, Harley sees the light in the moving clock traveling a longer path than Mo sees moving with the clock, which means that he sees her clock ticking slower than his, even though the two clocks are identical.

If we look at the path followed by the light in each of the two clocks, we can see this very clearly. Light in the stationary clock moves directly up and down as it bounces between the mirrors. Light in the moving clock, on the other hand, must follow a zigzag path: if the light leaving the bottom mirror went straight up, it would miss the top mirror as it moved rapidly from left to right. The light must move up and to the right in order to make it to the top mirror in time to be reflected back down. The zigzag path followed by the light in the moving clock is longer than the straight up-and-down path for the stationary clock, meaning that the time between ticks for the moving clock is longer, since all observers see exactly the same speed for light. Light moving on the zigzag path covers a longer path at the same speed, meaning that the interval between ticks must be longer.

If we look at half of the round-trip, we see that the light path forms the long side of a right triangle, as shown in Figure 3.1. This lets us be very precise about exactly how much time passes between ticks of the moving clock, using basic geometry. We know the lengths of all three sides of this triangle in terms of things that we measure, and we know the **Pythagorean theorem,** which gives us the relationship between the lengths.

"We do?"

"Sure we do. The Pythagorean theorem is one of the oldest mathematical formulae in existence. It says that the square of the hypotenuse of a right triangle is—"

"Pies are square!"

Figure 3.1.

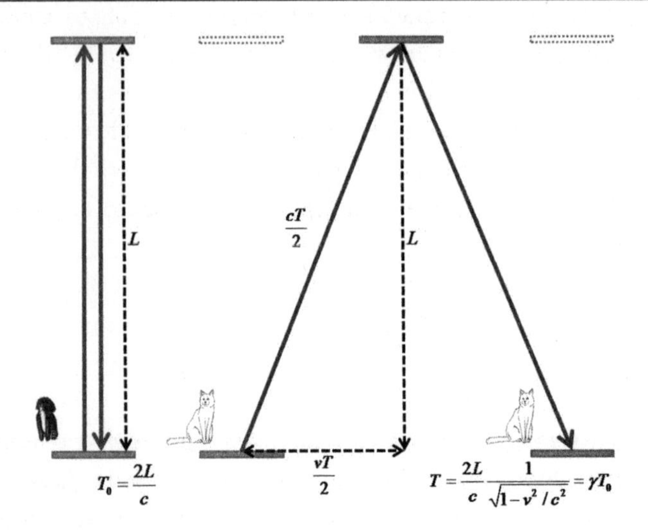

"What? No, there's no pi in this. Pi only turns up when you're talking about circles."

"And baking. I like square pie!"

"Yes, but we're talking about math. The Pythagorean theorem is $a^2 + b^2 = c^2$. The length of the long side, the hypotenuse, is c, and the lengths of the other two sides are a and b."

"Oh, *that* Pythagorean theorem . . ."

The vertical side of our light clock triangle is just the length of our clock. The horizontal side is the distance that the clock has moved in the time it takes light to get from the bottom mirror up to the top mirror, which is half the time T_{cat} between ticks of Mo's clock multiplied by her speed. The long side of the triangle is the distance that light travels in that same amount of time, namely, the speed of light c multiplied by T_{cat}. These lengths are related by the Pythagorean theorem, which lets us determine the time T_{cat} in terms of the length of our clock, the speed of light, and the speed of the moving clock. These are all things we know—the length is something we measured when we made the clock, the speed of light is a constant of nature, and the speed v is the speed of the cat going by (and any dog worth his kibble always knows *exactly* how fast a cat is moving)—which lets us get a value that we can compare to the time for a stationary clock.

When we go through the math, we find that the time between ticks of Mo's clock according to Harley is equal to the time between ticks of his clock multiplied by a number that depends on the speed of the moving clock and the speed of light:

$$T_{cat} = \frac{1}{\sqrt{1 - v^2 / c^2}} T_{dog} = \gamma T_{dog}$$

The **Lorentz factor** $\gamma = \frac{1}{\sqrt{1 - \frac{v^2}{c^2}}}$ appears over and over in relativity, so it's generally given its own symbol, a lowercase Greek letter gamma (γ), because physicists are lazy and don't want to write $\frac{1}{\sqrt{1 - \frac{v^2}{c^2}}}$ more than necessary. This factor tells us exactly how much slower a moving clock runs: for every tick of Mo's clock, Harley sees his own clock tick γ times.

"So, why does some stupid cat get to have time run slow, while the dog's clock runs the same as it always did?"

"That's what he sees when he watches Mo going by. Because Mo's moving relative to Harley, her clock is seen to run slow. In her frame, though, it's the other way around."

"Wait, doesn't the cat see the stationary clock running fast?"

"No, because according to Mo, her clock is running at the same rate it always did. In the cat's frame, Harley's the one who's moving, so it's his clock running slow."

"That doesn't make any sense, though."

"Yes, it does. Watch."

As strange as the idea of moving clocks running slowly may seem, that's only the beginning. To appreciate the real strangeness of relativity, we need to look at the same scenario from Mo's point of view as she goes by, looking into the lab at the dog. Whereas Harley saw himself as standing still and Mo as moving by from left to right, Mo sees herself as standing still, while Harley moves by from right to left.

The two situations are mirror images of each other, and as we see in Figure 3.2, the distances involved are exactly the same, so the conclusion is unchanged: the stationary clock ticks γ times for every tick of the moving clock. Only in this version, it's Harley whose clock is moving, and Mo's clock is stationary.

Looking at the light clock from both points of view demonstrates two critical ideas about relativity. First, it reinforces the idea that relative motion is all that matters. The ticking of the moving clock depends only on the speed of one relative to the other, not the direction of motion or any fixed objects. A third observer—say, a bunny running away from the cat at a slightly higher speed—would see both clocks running slow by different amounts: Harley's by an amount determined by his speed relative to the bunny, and Mo's by a smaller amount, determined by her speed relative to the bunny (as the bunny pulls away from Mo, she appears to have a small velocity in the opposite direction).

The symmetry between the dog's observation of the cat and the cat's observation of the dog also reinforces the principle of relativity: the laws of physics work exactly the same way for all observers moving at constant

Figure 3.2.

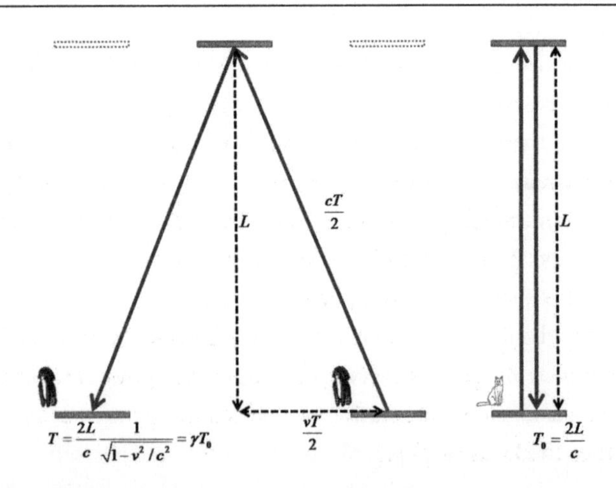

speed. If Mo saw Harley's clock moving at a different rate than he saw her clock running, then they could distinguish the stationary frame from the moving frame. The principle of relativity does not allow this, though; thus, Mo must see Harley's clock running slow by the same factor that he sees her clock running slow.

"Wouldn't it be easier if all their clocks ran at exactly the same rate?"

"Well, the math would be simpler, maybe, but that's not the way the world works. If all clocks everywhere ran at exactly the same rate, no matter how fast they were moving relative to other clocks, we'd need a completely different theory of electromagnetism, for example."

"Would that really be so bad?"

"Given that all of the atoms making up your body and mine and the food that you eat are ultimately held together by electromagnetic forces, yes, it would be kind of bad."

"Oh, all right. But how do we figure out who's right?"

"What do you mean, who's right?"

"Whose clock is really running slower? The dog's or the cat's?"

"Once again, you're both right, in your own frame of reference. There is no experiment either can do that will say anything other than that the other's clock is moving slow. As long as they're both moving with respect to each other, each will see the other's clock taking a longer time between ticks."

"I thought the clock just appeared to run slow? You know, because it's in a different frame?"

"It's often phrased that way, but I would say that the moving clock does run slow, because there is no measurement you can do that will give you any other result. Since reality is ultimately based on what we measure, I feel justified using 'is slow' rather than 'appears slow.'"

"I guess. It's awfully philosopher-ish, though."

"Maybe a bit. There's certainly a fine tradition of physicists dabbling in philosophy, and vice versa. Both philosophy and physics require you to be very careful about your definitions, and relativity requires a lot of care even by the standards of physics."

"See, dogs are much more laid-back. I figure that words mean whatever I need them to mean when I say them."

"How very Lewis Carroll. It does explain a lot, though."

"Thanks!"

The light clock is a simple argument that leads to a surprising conclusion. If moving clocks take a longer time between ticks than stationary clocks, why isn't this an everyday phenomenon? Why does this seem so surprising?

The answer has to do with the value of the Lorentz factor, which increases very slowly at low and moderate speeds but shoots up very dramatically as the speed approaches the speed of light. For any velocity we're likely to encounter in everyday life, though, γ is very close to 1. The speed of light is 299,792,458 m/s, while the speed of a jet airplane is only around 250 m/s. A moving light clock would need to be traveling at 14 percent of the speed of light, or 42,000,000 m/s, in order for the stationary clock to tick 101 times for 100 ticks of the moving clock.

The Lorentz factor increases dramatically at higher speeds, as we see in Figure 3.3. At 90 percent of the speed of light, it's 2.3. At 95 percent of the speed of light, it's 3.2. Increasing the speed of an object from 15 to 20 percent of the speed of light changes the value of γ by just 0.01, while the same 5 percent increase starting at 90 percent of the speed of light changes γ by ninety times as much. And the change keeps increasing: going from 95 to 99 percent of the speed of light more than doubles γ to 7.1, and at 99.9 percent of the speed of light, γ is 22.4. If we want to see a dramatic change in the ticking of a moving clock, then, we need to get it moving at very close to the speed of light.

"That's just because the light thingy is a stupid way to make a clock."

"What?"

"I mean, who wants a clock that ticks every 6 ns? It probably runs slow because it's a silly design, and if you made a more sensible clock, it wouldn't have this problem."

"That's the thing, though. Relativity requires that all clocks work at the same relative rate. You can build whatever sort of clock you like, and you'll

Figure 3.3.

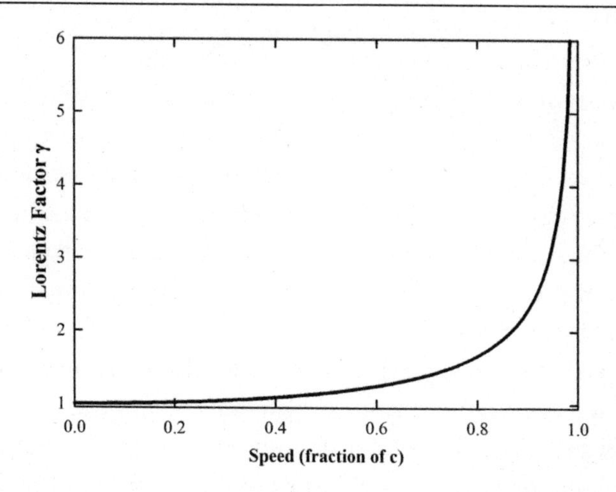

find the same thing. A moving clock ticks slower than a stationary clock. The ticking rate of any clock is ultimately determined by the laws of physics, and those have to be the same for all observers."

"Sure, for the light clock, but what about a pendulum clock?"

"If the pendulum swings along the direction of motion, it will appear to have a different speed as it swings forward than as it swings backward. The details are messy, but when you work it out, the pendulum takes a longer time to complete a round-trip according to Harley in the lab than it does for Mo moving with the clock."

"Well, what about your fancy atomic clocks?"

"The ticking of an atomic clock is based on the oscillation of light associated with an atomic transition. Those oscillations look different to people in different frames of reference; thus, a moving atomic clock runs slow."

"I don't know if I believe that . . ."

"Believe it. They've done the experiment, and it works out fine."

"Really? They've tested this?"

"Of course."

"Well, why didn't you say that? Enough *gedankenexperimenten*; tell me about the cool real experiments!"

TIME SLOWS WHEN YOU'RE FLYING: EXPERIMENTAL TIME DILATION

While the slowing effect of motion at ordinary speeds is extremely small, it can be measured, given precise enough measuring equipment. The most famous direct measurement of the effects of relativity on moving clocks was made by Joseph Hafele and Richard Keating in 1971. Hafele and Keating carried four portable atomic clocks on commercial airline flights around the world, then compared their clocks to the earthbound atomic clocks at the US Naval Observatory.

The Hafele and Keating experiment is complicated by the fact that, in addition to moving fast, airplanes fly high above the ground, and **general relativity** says that clocks at high altitude run fast (see Chapter 9). The rotational motion of the Earth is another complication, as the surface clocks are themselves moving, which must be accounted for.

Taking all these factors into consideration, Hafele and Keating predicted that clocks flown around the world in an easterly direction, taking about forty-one hours to complete the circuit, should run slow by 40 ± 23 ns, and when they did the experiment, the clocks were slow by 59 ± 10 ns. Westbound clocks were predicted to run fast by 275 ± 21 ns, and the experimental result was 273 ± 7 ns. In 1996, to commemorate the twenty-fifth anniversary of the Hafele and Keating experiment, the BBC arranged a partial repeat of the experiment, with a single, more modern atomic clock flown from Washington to London and back. The clock was predicted to run slow by 39.8 ns over the fourteen-hour flight time and was measured to be slow by 39 ± 2 ns, in excellent agreement with the prediction.

"Wait a minute—I thought you said you can't tell whose clock is moving slow? A dog on the ground should see the clock on the plane running slow, but shouldn't the cat on the plane see the dog's clock running slow?"

"While the plane is in motion, yes. But the flying clock experiment necessarily involves acceleration as well as motion at constant speed. The clocks start off at rest on the ground, then speed up as the plane takes off, and slow down as the plane lands. While they're speeding up and slowing down, the situation is more complicated."

"More complicated, how?"

"Well, when you speed up or slow down, you're not in an inertial frame anymore, so Newton's laws of motion don't appear to hold. It's like the soda on the dashboard back in Chapter 1—if you step on the gas with something sitting on the dashboard of your car, the car can accelerate out from under it. To someone in the car imagining himself to be at rest at all times, though, it looks like the object spontaneously leapt into motion with nothing pushing it."

"So, when the plane takes off, the clock lurches backwards, and that makes it run slow?"

"It's not quite that simple. The important thing, though, is that when the plane is accelerating, it's possible to tell which clock is which—the laws of physics don't quite look the same to a cat in the accelerating plane as to the dog on the ground. That lets you distinguish between the two observers, and when you go through all the details, the clock that accelerated and decelerated is the one that ends up running slow. Or fast, in the case of the westbound clock."

"Can't you just, you know, do it without accelerating?"

"Well, no, kind of by definition. 'Accelerate' just means 'change velocity,' and if you want the two clocks to start and end in the same place in the same frame, so you can compare them directly, at least one of them is going to have to accelerate at some point. You can increase or decrease the time spent accelerating, but you can't eliminate it."

"What if you don't go as fast?"

"If you don't go as fast, the effect gets much smaller, but you can still see it if you use good enough clocks."

Atomic clocks have improved dramatically since the time of Hafele and Keating, and even since 1996, enabling ever more precise tests of relativistic time effects. In 2010, scientists at the National Institute of Standards and Technology in Boulder, Colorado, demonstrated a new type of atomic clock based on a trapped aluminum ion. This clock is almost one hundred times more stable than the best cesium atomic clocks, which allowed them to do the most impressive measurement of relativistic effects to date.

The aluminum ion clocks are based on light absorbed by a single aluminum atom with one electron missing, which is suspended in space by using high-voltage electrodes a small distance away. In ordinary operation, the ion is held in place at the center of the trap, moving very little. With a slight modification to the trap, however, they can set the ion rocking back and forth at any speed they like.

To demonstrate the effects of relativity, they built two identical clocks and set one ion in motion while the other remained stationary. Then they compared the rates at which the two clocks ticked and measured the difference for several different velocities, from 0 to 40 m/s (about 90 mph).

As we see in Figure 3.4, they found a clear difference between the two clocks, even for ion velocities as low as 4 m/s, about the pace of a brisk walk. The difference increased as they increased the speed of the moving ion, exactly as predicted by relativity. This shows clearly and unequivocally that moving clocks run slow, even at low speeds.

"See! I told you I could slow time by running fast!"

"By a tiny, tiny amount. You need a state-of-the-art atomic clock to have any hope at all of detecting it."

"How tiny is tiny?"

"Well, the light they were using as a reference oscillated 1.2 quadrillion* times per second, and at their highest speed, the two clocks differed by less than ten oscillations per second."

"Oh. That's pretty tiny. Don't you have anything more . . . you know, dramatic?"

"Sure, but not involving anything as big as a dog or an atomic clock."

---

* One quadrillion equals 10^{15}, or 1,000,000,000,000,000.

Figure 3.4. Difference between the rates of a stationary ion clock and a moving ion clock for different speeds. The solid line shows the time dilation predicted by special relativity. (Modified from C. Chou et al., "Optical Clocks and Relativity," *Science* 329, no. 5999 [2010]: 1630–1633; used with permission.)

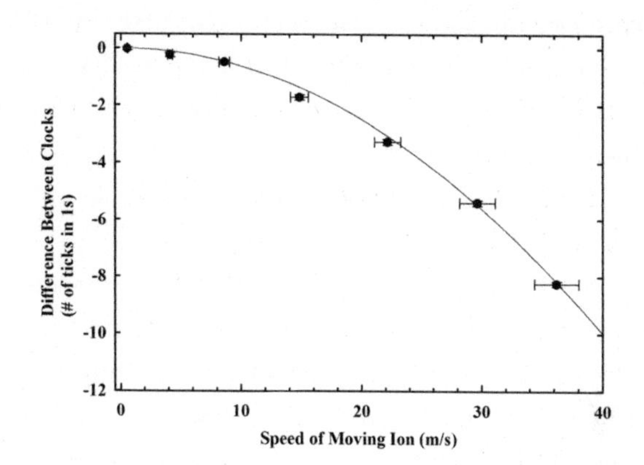

As impressive as the atomic clock tests are, it would be nice to see a demonstration of the more dramatic predictions of relativistic time dilation. Sadly, science has not yet found a way to accelerate a dog to a significant fraction of the speed of light (even Harley at his most excitable doesn't achieve those speeds), but we can see the effects of relativity on subatomic particles, which are easily accelerated to high speeds.

The most dramatic example of the clock-slowing effects of relativity involves exotic particles known as **muons.** A muon is a subatomic particle similar to an electron, but around two hundred times more massive.* Muons are unstable, with a lifetime of around 2.2 microseconds** (μs), after which they decay into an electron and a **neutrino.**

While muons can be created in the laboratory, they also occur in nature, such as when **cosmic rays** strike the upper atmosphere. A cosmic ray particle striking a molecule in the upper atmosphere will create a "shower" of many different kinds of particles, including some muons, which are created moving at better than 99 percent of the speed of light.

---

* We'll talk more about muons in Chapter 8.

** A microsecond equals one millionth of a second, so in this instance, 2.2 μs = 0.0000022 seconds.

Of course, even at 99 percent of the speed of light, you can't get very far in 2.2 μs—at that speed, a muon should travel only about 660m. Thus, you would not expect muons created at an altitude of 10 km to have any chance of reaching the ground—the trip would take them 34 μs, a little more than fifteen times the average muon lifetime. If you looked at 10 million muons with this speed starting at an altitude of 10 km, you would expect only one or two to reach sea level.

In reality, though, we detect substantial numbers of muons at sea level—in fact, sensitive particle astrophysics experiments need to be carried out deep underground in order to avoid detecting muons created by cosmic rays.* How is this possible? The answer is relativity: at 99 percent of the speed of light, the Lorentz factor, γ, is 7.09, which means that a muon moving at 99 percent of the speed of light has its internal clock slowed by that factor. A single 2.2 μs lifetime according to the fast-moving muon lasts 15.6 μs according to an observer on the ground. In that case, more than 10 percent of the muons should survive to reach the ground, which is exactly what we see experimentally. The same effect is seen with fast-moving muons in particle physics experiments: muons moving at speeds close to the speed of light last longer before decaying than muons at rest, exactly as predicted by relativity.

So, as strange as it may seem at first glance, the slowing of moving clocks predicted by relativity is a well-confirmed effect. We can measure the tiny shifts predicted for ordinary speeds using sensitive atomic clocks, and we can also see the dramatic effects predicted for fast-moving objects by looking at the lifetime of speedy subatomic particles. Moving clocks, no matter how they're made, run slow, exactly as predicted by Einstein.

"That's pretty cool, dude, but I think there's a problem."

---

* The world's best detector of the extremely light and difficult-to-detect neutrino, the Sudbury Neutrino Observatory, is located 6,800 ft. underground in a nickel mine in Sudbury, Ontario, because more than a mile of solid rock is needed to prevent cosmic ray particles from reaching the detector. It's easier for particle astrophysicists to put their detectors at the bottom of a working mine, with underground blasting taking place nearby, than to sift out a tiny number of neutrinos from the vast stream of cosmic-ray particles reaching ground level.

"Yeah? What problem."

"Well, the muons last 15 μs according to people on the ground, right?"

"Yes."

"But according to the muons themselves, they only last for 2.2 μs. I mean, that's what the slow ticking would mean, right? So, 15 μs goes by on a clock on the ground, but it's only 2.2 μs on a clock riding with the muon."

"Exactly."

"But you said yourself that they should only go about 660m during that time. So, how can a muon see itself reach the ground? Didn't you do something wrong?"

"No, that's exactly right. That's the next weird prediction from relativity—according to the muon, the distance to the ground is not 10 km but just 1.4 km."

"OK, what? That's just bizarre."

"Bizarre, but true. Moving objects shrink in relativity, and that's the subject of the next chapter . . ."

Chapter 4

HONEY, I SHRANK THE BUNNIES: LENGTH CONTRACTION

EMMY'S PATROLLING THE BACKYARD, and when I head out to see what she's up to, she explodes off the patio and runs in a huge circle to the back of the yard, then back close to the house, then along the neighbors' fence. She pulls up short at the fence, then takes off again, even faster, for a second pass.

"What in the world are you doing?" I ask. She skids to a halt at my feet, panting.

"I've figured out how to make it through the fence," she says. "I'm going to squeeze through the gaps and get those bunnies in the next yard."

"I don't think you're going to make it through the bars," I say. "You're significantly wider than the spacing between the bars of the fence."

"That's because I'm standing still," she says cheerfully. "Once I get going fast enough, I'll fit. Because of physics!"

"This is another relativity-based scheme, isn't it?" I sigh. "All right, what's the latest brilliant plan?"

"Well, according to relativity, moving objects shrink, right? So if I go fast enough, I'll be small enough to fit through the bars!"

"Running really fast might make you thinner if you exercise enough to lose some weight, but it's not going to make you thin enough to fit through the bars. While fast-moving objects in relativity do get smaller, they only shrink along the direction of motion. Running really, really fast might contract the distance between your nose and your tail, according to somebody in the yard, but it won't make you thin enough to squeeze between the bars."

"Of course not, silly," she says. "That's why I'm running *along* the fence. If I run really fast, almost parallel to the fence, I'll be shorter than the spacing between the bars, and I can just squeak through at an angle. It'll be a little tricky, but I'm very agile."

"OK, but there's still a problem. What matters for length contraction is the velocity relative to the observer. An observer standing next to the fence will see you get shorter, but *you* will see the fence get shorter. The faster you go, the smaller the space between the bars will get. Your brilliant plan will have exactly the opposite effect from what you want."

She looks puzzled. "No it won't," she says, brow furrowed. "I'm the one moving, so I get smaller. The fence isn't moving, so it stays the same."

"It's relativity, remember. What matters is the relative speed. An observer standing by the fence sees the fence standing still and you moving, but an observer moving along with you would see you standing still and the fence moving."

"So, wait, an observer in the yard sees me get shorter?"

"Right."

"And I see the fence get shorter?"

"Exactly."

"But that doesn't make any sense! Both of those can't be true at the same time, can they?"

"They can, and they are."

"So, what, everything shrinks when I move really fast?"

"Each observer sees the other's object shrink, but they don't agree with each other's measurements. The key to understanding it is to think about what it really means to measure the length of something. What's the definition of length?"

"What do you mean, 'What's the definition of length?' It's the distance between two points. My length is the distance from the tip of my excellent nose to the end of my very nice tail."

"Ah, but you've left out a key qualifier—another of those hidden assumptions you don't realize you're making. Your length is the distance between your nose and tail *measured at the same time.*"

"Well, yeah. That's obvious."

"Right. And that's the solution to the problem—the timing of the measurements."

"Oh, OK. I see." She thinks for a minute. "No I don't."

"Well, we already said that moving observers disagree about the timing of events, right?"

"Sure."

"So, events that one observer sees as happening simultaneously—measuring the distance from your nose to your tail, say—a different observer sees happening at different times."

"And because I'm a very fast dog, I've moved between measurements, so the length is different."

"Exactly. When you work out the timing, the distance you move between measurements accounts for the discrepancy."

"So which of us is really right?"

"You're both right, in your own individual frames of reference. There is no measurement anyone else can do that will yield any value other than your contracted length. And there's no measurement you can do that will show anything other than the contracted spacing between the bars of the fence."

"Oh, OK." She thinks for a minute. "You know what?"

"What?"

"Relativity is really weird."

"You got that right." I scratch behind her ears. "The important thing, though, is that it's not going to get you through the fence to the neighbors' yard, no matter how fast you run. And, anyway, you couldn't possibly run fast enough for relativity to be a factor."

"Oh, yeah? I'm very fast!"

"You're very fast, but in order for a stationary observer to see your length contracted to the 10 cm spacing between the bars of the fence, you'd need to be moving at something like 99.6 percent of the speed of light. That's a hair under 299,000,000 m/s."

"Oh. You're right. I'm not that fast."

"No, you're not."

"Not yet, anyway. But if I was, I could really catch some bunnies!"

"I bet you could."

"So I better get back to my sprint training!" And with that, she turns around and resumes running full out in a huge circle in the backyard. I just shake my head and go back inside.

At the end of last chapter, we looked at the decay of muons created high in Earth's atmosphere as a dramatic example of relativistic time dilation: because of the particles' high speed, a single 2.2 μs muon lifetime stretches to 15.6 μs, giving them enough time to cover the distance between their creation point and sea level. Although this allows a stationary observer to explain the detection of muons at low altitudes, it creates a problem for explaining the events in the muons' frame of reference.

An observer traveling with the muon will necessarily have a clock that ticks at the same rate as the internal "clock" determining the muon's decay. To that observer, the muon lifetime remains merely 2.2 μs. This, in turn, means that in order for the muon to be detected at low altitude, the distance must be smaller than that seen by a stationary observer.

This length contraction strikes humans as one of the stranger consequences of relativity. Surprisingly, it was also one of the first relativistic effects predicted historically by George FitzGerald and Hendrik Lorentz.

The idea proved a little too strange for the physicists of the 1890s to accept, though, and it wasn't until Einstein showed that length contraction is a natural consequence of the relativity of time that the idea became part of modern physics.

In this chapter, we talk about the historical origin of length contraction in an attempt to explain the Michelson-Morley experiment and how length contraction was eventually explained by Einstein. Just as the explanation of time dilation in the previous chapter required us to reexamine the concept of time at a fundamental level, understanding the physics of length contraction requires us to reexamine the idea of space and how we measure it.

TAKE THE LONG WAY HOME: MICHELSON-MORLEY REVISITED

The idea that a moving object should shrink along the direction of motion was first proposed by Irish physicist George FitzGerald in 1889, very shortly after Albert Michelson and Edward Morley's experiment. Dutch physicist Hendrik Lorentz arrived at the same idea independently a couple years later and worked out more of the mathematical details. Both Lorentz and FitzGerald realized that contraction along the direction of motion was necessary to explain the Michelson-Morley result, even assuming the speed of light was constant.

To understand the need for contraction, let's look at a moving Michelson interferometer in more detail. An observer at rest relative to the interferometer sees exactly the picture we drew in Chapter 2, with light beams taking right-angle paths. A moving observer, however, sees the light taking a very different path. The light reflected perpendicular to the motion travels along an angled path, just as in the light clock. The light transmitted along the direction of motion stays on the same axis, but it too travels a different distance from that seen by the stationary observer because the beam splitter and mirror move during the time the light is in flight (see Figure 4.1).

Figure 4.1. Left: A Michelson interferometer seen by an observer who is stationary relative to the interferometer. Right: The interferometer as seen by an observer moving from right to left relative to the interferometer, showing the observed path of the light beams.

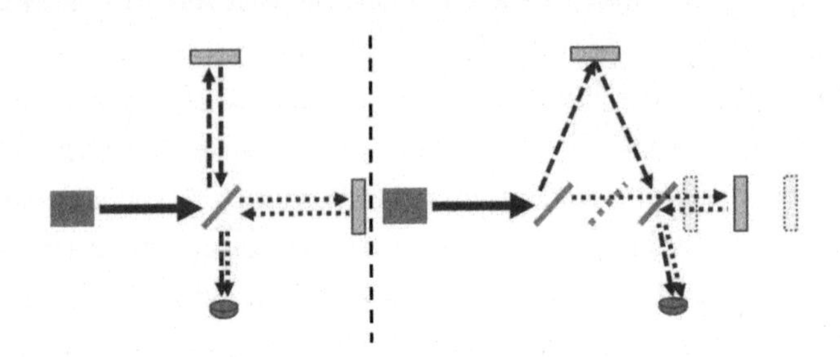

The distance traveled along the perpendicular arm is identical to that seen in the light clock in the last chapter: a moving observer looking at the interferometer (or a stationary observer looking at a moving interferometer) will see the light follow a path that is longer than the stationary observer's path by a factor of γ. To give a concrete example, an observer moving at one-quarter the speed of light looking at an interferometer with arms 1m in length would see light travel 2.066m, an additional 6.6 cm above the distance seen by the stationary observer.

Finding the distance traveled by light moving along the direction of motion is more complicated, but we can understand it by zooming in to look at just that arm of the interferometer. Figure 4.2 shows the important steps for an interferometer as seen by an observer moving at one-quarter the speed of light from right to left.

A pulse of light leaving the beam splitter would have to travel 1m to reach the mirror, if the interferometer were stationary. According to a moving observer, though, in the time it takes the light to reach the mirror's initial position, the mirror has moved away from the light by some distance. When the light catches up to the mirror and reflects off it, the total distance traveled is 1.333m.

On the return trip, the beam splitter is moving *toward* the light, and thus the distance traveled is shorter. When the light gets back to the beam splitter, it has traveled only 0.8m from where it was when it hit the mirror. The total distance traveled is 1.333m + 0.800m = 2.133m. In terms of the Lorentz factor, this is a factor of γ^2 longer than the path seen by a stationary observer.

This presents a problem for any attempt to explain the Michelson-Morley result using a constant speed of light for all observers. Using the same sort of reasoning we employed in considering a moving light clock, we have found that light along the perpendicular arm travels a different distance than light along the parallel arm. The effect is small for low velocities but should've been easily detectable when rotating the interferometer.

FitzGerald and Lorentz realized, however, that the problem could be fixed if the moving observer sees a shorter distance between the beam splitter and mirror along the parallel arm. Using the numbers from our example above, if the parallel arm was only 0.968m long—that is, the length according to a stationary observer divided by the Lorentz factor, γ—the distances traveled along each arm would be exactly equal, which would explain the Michelson-Morley result.

Figure 4.2. The important steps for light traveling through a Michelson interferometer moving at one-quarter the speed of light. When the light has covered the 1m length of the arm, the mirror has moved away; by the time the light catches up, it has covered 1.333m. On the return trip, the beam splitter catches up to the light, and it covers only 0.8m. The total distance traveled is 2.133m, rather than the 2.000m for a stationary interferometer.

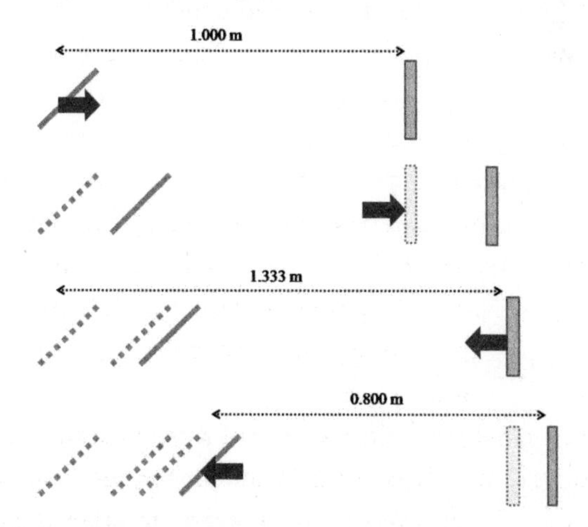

This is an elegant solution to the problem, but in the early 1890s, there was no convincing reason why moving objects ought to shrink. FitzGerald and Lorentz attempted to explain it through a modification of the electromagnetic forces holding material objects together, but that struck most physicists of the day as an ad hoc justification rather than a convincing argument. Thus, while the mathematical method undeniably gave the right answer, nobody really believed it until Einstein's relativity paper of 1905.

"I can't say I blame them. I mean, that's just crazy talk."

"It sounds strange, but you have to admit, it does fix the problem with the Michelson-Morley result."

"Sure, but if you're going to start changing the sizes of things, why not just stretch the perpendicular arm out? That'd fix the problem, too."

"I guess, but why would that be any more acceptable than shrinking along the direction of motion?"

"Well, objects that start moving stretch out all the time. I've seen it on TV."

"Cartoons don't count as evidence. Anyway, the direction perpendicular to the motion can't change, because that would mess up the time dilation formula, which we've already seen works really well. Any change that occurs can only be along the direction of the motion."

"Yeah, but it doesn't make any sense. There's no reason why moving objects should shrink."

"Actually, it makes perfect sense that measurements along the direction of motion should change—provided, of course, that you think very carefully about what it means to measure the length of an object."

TIMING IS EVERYTHING: LENGTH MEASUREMENTS AND CLOCK SYNCHRONIZATION

Just as we needed to look at the working of a clock in order to understand the effects of relativity on time, understanding the effects of relativity on space requires us to think carefully about measuring distances. In particular, we need to think carefully about how we define the length of an object.

This is another question that may seem almost too obvious to worry about—the length of a dog is the distance between the tip of her nose and the end of her tail, and that's that. There's an unspoken qualifier there, though: in order to accurately measure the length of a dog, you need to measure the distance from the tip of her nose to the end of her tail *at the same instant in time*. This is almost never an issue for ordinary dogs at ordinary speeds, so we don't even need to think about it—the measurement process is effortless, requiring little more than a glance. When we start to consider objects that are moving at very high speeds, though, the question of timing becomes problematic: since observers moving relative to one another see their clocks run at different speeds, making sure that the measurements of the positions of the dog's tail and nose are simultaneous becomes problematic.

The issue involved is subtle but crucial to relativity. The problem of length measurement is really a problem of **synchronization:** to measure the length of a dog, you measure the position of her nose at some instant in time, say, by making a chalk mark on the floor, and at the same instant, you measure the position of her tail, by making a second chalk mark. The process of recording each position is necessarily done at that particular position, which means that a proper measurement of the dog's length is a job requiring *two* observers, one at the nose and one at the tail.* These two observers must have synchronized watches to enable them to make their marks at the same instant, and this is where the problem comes in. In relativity, it is impossible for two clocks in different locations to be synchronized in *all* frames of reference. Two observers at rest relative to one another may agree that their watches are synchronized, but a third observer moving relative to the first two will see not only both of their clocks as running slowly but one clock as ahead of the other by an amount that depends on their separation.

---

* An ambidextrous human might be able to manage simultaneous marks at the head and tail of a smallish dog by making one mark with each hand, but as a general matter, you need two observers to do the job right.

Figure 4.3.

We can make sense of this by thinking about measurements on a larger scale and devising a process to synchronize two distant clocks. Let's imagine, for example, that Emmy the physics dog would like to know the precise width of her yard, so as to know the proper length of fence needed to keep annoying cats out. We can accomplish this measurement by stationing one human at either end of the yard, with a measuring tape between them, and asking them to record the position along the tape that corresponds to their edge of the yard. To ensure that they make the measurement at exactly the same instant, we can flash a bright light at the center of the yard, halfway between them, and have them record their measurement at the instant the light from the flash reaches them (see Figure 4.3).*

This may seem overly elaborate for fence-purchasing purposes, but even the pickiest human physicist will agree that this method ensures a properly simultaneous measurement of the edges of the yard, thus its length—provided, of course, that the human physicist is not moving relative to Emmy and her surveyors. From the point of view of a moving observer, say, Nero the cat casing the neighborhood for dogs to annoy, the process looks very different (see Figure 4.4).

In Nero's frame of reference, he is stationary, while Emmy, her human assistants, and even her yard are moving. When the flash of light goes off,

* They'll need very good reflexes for this—a suburban dog on a quarter-acre lot would have just over 100 ft. of property line on any side, so the light from the flash would take just over 50 ns to reach the surveyors.

Figure 4.4.

Nero sees the lead human running away from the light, while the human bringing up the rear catches up to the light pulse. As a result, the light reaches the back human before the front human. In Nero's frame, then, the measurements that Emmy and her humans agreed were perfectly simultaneous are flawed: the rear of the (moving) yard was measured first, and the front of the yard was measured later. In between the two measurements, the yard has moved, leading Emmy and her humans to think that the yard is wider than its width according to Nero.

"That's just like a cat, too."

"What is?"

"You know, claiming that my perfectly simultaneous yard-measuring procedure wasn't done properly. I had humans doing the measuring and everything."

"That's the whole point, though. In Nero's frame of reference, he's absolutely correct. He quite clearly sees the rear measurement made first, while the front measurement is made later, giving an inflated length. Nero, on the other hand, sees your yard as being shorter than you do."

"What do I see, when the sneaky cat makes his measurements, then?"

"Basically the reverse process. If Nero enlists human minions to measure your yard, you see the front measurement made first, while the back measurement is made later, after it's moved closer to the first measured position."

"Again, just like a cat. They're always using this kind of underhanded trickery. You can't trust them at all."

"Relativity doesn't depend on the duplicity of cats, though. The species making the measurements doesn't matter—no matter what procedure you use to ensure properly simultaneous length measurements in your frame, the measurements will occur at different times according to any observer who is moving relative to you, and vice versa. Events that are simultaneous in one frame take place at different times in any frame moving relative to that one."

"Well, okay, I guess. But I still say cats aren't to be trusted. Particularly not cats named after insane Romans."

"I'll try to keep that in mind."

As a general rule, moving observers looking at events that are simultaneous in some other frame see the events in the direction of their motion happen before events in the direction opposite the motion. In Emmy's frame, Nero is moving from right to left; thus, he sees Emmy's left-hand measurement made first. In Nero's frame, Emmy is moving from left to right; thus, she sees Nero's right-hand measurement made first. This difference in timing explains the different distances measured in different frames.

Deriving the mathematical formula for relativistic length contraction is more complicated than the earlier discussion of time dilation using the light clock experiment,* but the end result is clear from the Michelson interferometer: the distance shrinks by the Lorentz factor, γ. We can also see this from the muon example of the last chapter: muons with a 2.2 μs lifetime created at an altitude of 10 km moving at 99 percent of the speed of light will last 15.6 μs according to an observer on the ground, giving them enough time to reach sea level. From the point of view of an observer moving along with the muons, though, they are at rest, and their lifetime remains unchanged. Both observers must agree, though, that some muons end up at sea level, which means the distance between the high-altitude

* The basic approach is the same as in the Michelson interferometer discussion above.

point where they are created and the sea level must be shorter by the same Lorentz factor, γ, by which their clocks run slow. Any other result would run counter to the principle of relativity, requiring physics in the muon's frame to appear radically different from physics in the Earth's frame.

This muon example also provides the first indication of the mingling of space and time in relativity. Unlike classical physics, in which space is space, time is time, and that's that, in relativity, their roles are more fluid. What one observer sees as a contraction of space—the length of a moving object shrinking—another observer sees as a stretching of time—a moving clock running slow. The synchronization of clocks is another aspect of this mingling of space and time—observers at different places will see events happening at different times. We explore the combined nature of **spacetime** further in the next chapter.

YOU CAN'T TELL THE OBSERVERS WITHOUT A PROGRAM: RESOLVING PARADOXES

The issue of clock synchronization and the timing of measurements is the key to understanding most of the "paradoxes" of relativity. Many discussions of relativity will include one or more hypothetical situations in which seemingly impossible things appear to take place, causing great confusion among humans reading about the subject, to say nothing of dogs. These apparent paradoxes are almost always resolved by keeping careful track of exactly who makes the measurements, and how, at every step of the problem.

A classic example is the barn-and-pole paradox, which involves a running human carrying a pole through a barn. As this scenario strikes even humans as kind of daft, though, let's consider a more sensible dog-and-leash paradox.

Imagine a fast-moving dog, say, Anson the Lab mix running through a park, having shaken loose from his humans. He's towing a leash that is 3m in length (about 10 ft.), and the park is full of slow-moving humans who would like to catch him and whom he would prefer to avoid. At some point in his romp through the park, he dashes under an ornamental bridge that is 2m wide.

"Wait, why does he get to run around loose? This is my book. I should be the one getting to have fun."

"Yes, but you're a very good dog and would never slip loose from me while out on a walk. Would you?"

"Oh, yeah. Good point. I'd drag you along with me."

"Right, and I wouldn't fit under the bridge, so we need a different dog for this, OK?"

"OK."

Now, a 3m leash won't fit under a 2m-wide bridge, but assuming Anson is moving very fast, relativity tells us that at about three-quarters the speed of light, the leash shrinks to only 2m in length and thus can be completely contained under the bridge. Two humans trying to catch him, one trying to grab the leash before it disappears under the bridge and the other trying to snag his collar as he comes out from under the bridge, would say that, at some instant, the entire leash fits beneath the 2m bridge.

Of course, Anson is moving at the same speed as the leash, so in the unlikely event that he pays any attention to it, he still sees the leash as 3m in length. Anson sees the width of the bridge shrink, instead, since it's moving relative to him, so he sees the bridge as just 1.3m wide.

"Wait, you have a 3m leash fitting under a 1.3m bridge? What is this, M. C. Escher Park?"

"See, you're falling into the trap of this paradox."

"Well, yeah. I mean, if something that's 3m long fits in a space 1.3m long, that's quite the trap."

"It only seems that way, because you've lost track of who sees what. No observer sees both a 3m leash and a 1.3m bridge. Anson sees the bridge being short, and the leash long, while his humans see the bridge at its usual length and the leash as only 2m long."

"Yeah, but how does Anson understand the leash fitting under the bridge?"

"He doesn't, because he never sees the leash fitting under the bridge. Watch."

The "paradox" here has to do with the notion of fitting a 3m leash under a 1.3m bridge (without coiling it up). This seems like a problem, but it arises from confusion about who's observing what. In the correct relativistic description of events, no observer sees both the leash fitting under the bridge and the bridge being shortened. The humans see the leash shrink to 2m and fit under the bridge, while Anson sees the bridge shrink but never sees the leash fit under it.

So, what does Anson see? He sees his humans' measurement of the length being made incorrectly, at two different times. Where his humans see the front of the leash reach the front of the bridge at the same instant that the back of the leash passes under the back of the bridge, he sees those two events happen at different times. Anson's head reaches the front of the bridge about 7.5 ns before the back of the leash reaches the back of the bridge. When the front of the leash reaches the front of the bridge, 1.7m of leash are trailing behind, outside the bridge. The back of the leash doesn't reach the back of the bridge for another 7.5 ns, at which point the front of the leash is 1.7m past the front of the bridge, as Anson bounds through the park at three-quarters the speed of light.

There are a number of other variations on this sort of paradox, but most of them come down to the same sort of problem. When trying to understand the effects of relativity, it's critical to keep track of which observers make which measurements at which times. If you attribute the right measurement to the wrong observer, you will tie yourself in knots trying to figure out what's going on.

"OK, I get it now. The fast-moving dog sees his leash at the right size, but his humans following behind see it short because it takes light a while to get to them. Right?"

"No, no—this has nothing to do with the time it takes light to travel to an observer. If you try to include that, everything gets way more complicated, because you'll see light from different parts of the object at different times.

An object coming toward you might even appear elongated rather than contracted, because the light coming from the back end of the object takes longer to get to you, and so you'll see it later than the light coming from the front end. Length contraction assumes a different kind of measurement."

"What kind of measurement?"

"Basically, a network of observers at key positions, all with carefully synchronized clocks, record the instant an object of interest passes their positions. Then, at a later time, they compile all their observations and use them to reconstruct the series of events."

"That seems awfully complicated, dude. Why not just use your eyes?"

"You can do that and just record when you see objects passing key points, but it doesn't get you out of the need to reconstruct things after the fact. For example, if you see two flashes of light reach you simultaneously, does that mean the two flashes were simultaneous?"

"Well, yeah. Why wouldn't it?"

"Well, what if one of the flashes of light comes from a lamp in the next room, while the other is the sun glinting off a satellite up in orbit? In that case, the two actual flashes take place at two very different times. The light reflected off the satellite about a millisecond before the lamp turned on in the next room. The light from each reached you at the same time because the satellite is vastly farther away from you than the lamp is."

"So, it's like astronomy?"

"What?"

"You know, the way people on TV science shows always say that looking up at the stars is like looking back in time, because it takes light many years to travel to us."

"Oh, yeah. Trying to measure relativistic objects from a single position is sort of like that. In the same way that astronomers see distant galaxies the way they were millions of years ago, a single observer watching an object moving at relativistic speeds sees it where it was when the light left—which can be a substantial distance away from where the object *is* at the time of observation, if it's moving fast enough."

"So we see moving clocks running slow because we're seeing them as they were in the past?"

"No, relativistic shifts in clocks are what we see even after we've corrected for the speed of light delays. Since we know the speed of light, we can work backwards to figure out the correct position of a fast-moving object at various times from a set of pictures of the object from any given observation point. When we do that, we still find that the clock in the moving frame is running slow, and the moving object is shorter."

"I dunno. That's awfully indirect. Hey, are there any cool experiments demonstrating length contraction?"

"Not directly, no, because nobody has ever accelerated an object of substantial size up to the necessary speeds. You can do time dilation experiments with single atoms, because a single atom can act like a clock, but nothing depends on the size of a single atom in ways we can measure with the necessary precision."

"So, how do we know it's true, then?"

"Well, because we have done experiments that demonstrate the relativistic effects on time, and length contraction follows logically from time dilation. Since the time effects show up in experiments, we know that the length contraction effects have to be real as well."

"Wouldn't it be better to just accelerate a bunny to a speed close to the speed of light and measure its length?"

"It'd be nice to see the effect directly, but there's no good way to do it. It's extremely difficult to get macroscopic objects up to relativistic speeds, as we'll discuss in Chapter 6."

"I still think it'd be worth doing. It'd be way more fun to measure a relativistic bunny than to do boring logical inference from experiments with atomic clocks."

"I'll keep that in mind in case I ever end up with the billions of dollars it would take to build a bunny accelerator."

The notion that moving objects change their size is an extremely weird idea for most humans to get their heads around. The difficulty of reconciling this with our everyday view of moving objects is one of the main factors that delayed the acceptance of the theory until after Einstein published his 1905 papers. For the fifteen years before that, most physicists dismissed length contraction as an ad hoc justification for a mathematical anomaly.

The full understanding of length contraction came only after an understanding of the meaning of time, the effect of motion on its measurement, and the notion of **simultaneity**. Two events at different positions that one dog sees happening at the same time take place at different times according to any other observer—dog, cat, or bunny—moving relative to the original dog. Once we understand that, we can explain length contraction in terms of the timing of measurements, and we can see it not as some bizarre glitch in the equations but as an inevitable consequence of the way our universe works.*

Strange as they seem the first time you encounter them, these ideas of simultaneity and length contraction are only the beginning. In particular, they are the first indication of the merging of space and time, which is arguably the most dramatic consequence of relativity and the key to Einstein's greatest triumph, the general theory of relativity, which we will talk about in Chapter 9.

"Hang on—Chapter 9? This is only Chapter 4. If this timing business explains everything, how are you going to fill the next four chapters—with cloying stories about puppies and fluffy bunnies?"

"Oh, there's plenty to talk about before we get to general relativity—the notion of unified space and time, for one thing. And, of course, there's $E = mc^2$."

"That's right—you haven't explained yet how to convert squirrels into energy. That's my favorite part of relativity."

"The process by which *you* would convert squirrels into energy is more about biology than physics. Converting mass into energy in the $E = mc^2$ sense is a much more complicated process and not something that will be done to a squirrel any time soon."

---

* Even with the right view of time, it's not an easy step to make, though. Henri Poincaré had the time aspects of relativity worked out by 1904 but continued to think that length contraction needed to be a separate postulate of the theory even after Einstein's work was published.

"That's just because human physicists have their priorities out of whack. Let a really smart dog like me work on the problem, and we'll be flying around in squirrel-powered spaceships in no time."

"I'll believe it when I see it. We need to talk about space and time some more first, OK?"

"I guess so. As long as we get to the squirrel-powered spaceships soon."

Chapter 5

INTERVALS AND DIAGRAMS: AN INTRODUCTION TO SPACETIME

EMMY IS DOING HER BEST STEALTHY CREEP across the yard as I come outside, working her way toward a pair of mourning doves eating birdseed off the ground beneath the feeder. They get spooked by the sound of the screen door banging shut behind me and take off in a flurry of feathers and squawks.

Emmy makes a token run at them, but they're off the ground and out of reach before she gets anywhere near them. She sniffs around a little, then glares accusingly at me. "I was going to catch those bunnies, until you scared them off."

"First of all, you've stalked hundreds of animals that way, and not once have you caught anything. More importantly, though, those were birds, not bunnies."

"Maybe they looked like birds to you, but in my frame of reference, they were bunnies."

"OK, that makes even less sense than usual. What are you talking about?"

"It's simple relativity, dude. I was moving relative to you, so my measurements of the yard and its contents were different from yours."

"And how does that turn birds into bunnies?"

"Well, moving observers disagree about what time it is, right?"

"Yeah."

"And they disagree about the positions and sizes of objects, right?"

"Yeah."

"Right. So, those are such fundamental differences that I figure all bets are off, and anything goes. So, if I move at the right speed, birds become bunnies, and I can catch them without them flying away."

"Yeah . . . not so much. Relativity doesn't work that way."

"But it's all apocalyptic and stuff. Time is different; space is different. Things fall apart, the center cannot hold, mere anarchy is loosed upon the world . . ." She trails off, looking really distressed.

"You must really be bothered by this, if you're quoting Yeats."

"It's all freaky and makes my head hurt. All the stuff I thought was fixed and unchanging is all different."

I scratch behind her ears and sit down next to her. "It's not that bad. Relativity isn't a free-for-all; it's a rigorous scientific theory, making predictions within very tight limits. And while it changes some things that you might've thought were fundamental, it leaves a lot of other things unchanged."

"Like what?"

"Well, all observers agree about the speed of light, for one thing."

"Yeah, but that's the root cause of all the weird stuff. That's not exactly comforting."

"They also agree about their velocity relative to each other and about the simultaneity of events that occur at the same place."

"That's pretty weak. I mean, it's not much of a consensus reality to hang on to, you know? I mean, if you can't agree on the length of an object, what can you trust? Once you throw length into doubt, it's just a few short steps to blood-dimmed tides drowning the ceremony of innocence."

"Seriously, lay off the Yeats. It's true that different observers will disagree about the distance in space between two events and about their separation in time, but they do agree about the **spacetime interval** between them. That's constant for everyone."

"The what?"

"The spacetime interval. It's basically like the distance between two objects, when you treat time as just another dimension, like space."

"What does that mean?"

"Well, you know how you can figure the distance between two points by measuring the distance along two parallel axes, right? If an object is three steps east of you and four steps north, then it's five steps away in a direction 37 degrees east of north. That total distance of five steps will be the same, no matter how you orient your coordinates."

"Yeah, but that's just the problem. Relativity says that distance is different for observers moving relative to one another."

"Right, but if you generalize that distance to include the time between two events as well as their separation in space, then you get a number that doesn't change as you go from one frame to another."

"What do you mean?"

"Well, imagine you have two events that happen at the same time, according to one dog, at positions that are separated by 5m. According to that dog, the spacetime interval between those points is just that 5m: 5m in space and zero seconds in time."

"Yeah, so?"

"Now, think about what a moving cat going by at four-fifths the speed of light would see. According to that cat, the two events are separated by 8.3m in space, but one of them takes place 20 ns before the other. When you combine those numbers to get the spacetime interval between the two events, you end up with the same 5m of separation."

"Ummm . . . OK, but what does that mean?"

"Well, it means that space and time, which we think of as two very different dimensions, are just different aspects of the same thing. Events one observer sees separated only in space, another observer may see separated only in time. The total separation in space and time is the same for all observers, though. Which means that relativity isn't total anarchy—if you just shift what you're looking at a little bit, it's every bit as orderly as classical physics."

"OK, I guess that does sound a little better. So, this combination of space and time, what do you call it?"

"Ummm . . . spacetime."

"Spacetime? With a hyphen or without?"

"Either, really. Most physicists these days do without."

"You know, physics is cool, but physicists really stink at naming things. You should hire some English majors to help with that."

"Look, a lousy name is a small price to pay for avoiding a Yeatsian apocalypse, isn't it?"

"I suppose. I still think it needs a better name, though."

"You think about it, and let me know what you come up with. Meanwhile, if you see any rough beasts slouching around here, run them off, OK?"

"You can count on me!" She wags her tail happily and trots off to patrol along the fence.

Humans and dogs encountering relativity for the first time often come away thinking that it replaces the orderly world of classical physics with utter anarchy. On hearing that different observers disagree about the passage of time and the size of objects, they jump to the conclusion that absolutely everything is up for grabs in relativity. Some of them will reject the theory entirely, despite the massive amounts of evidence confirming its predictions.*

---

* As with most other forms of craziness, the Internet offers a wide array of sites denouncing relativity for one reason or another, often because they confuse relativity with relativism, the philosophical notion that cultural and moral questions can only be answered relative to some local standard, a cartoon version of which is often held to be responsible for all society's ills.

Nothing could be further from the reality of relativity than this picture of anarchy. Einstein himself never viewed relativity as a radical overthrow of existing physics; rather, he felt he was shoring up its foundations by properly incorporating the limitations of real measurements. He would have found the idea of an "anything-goes" universe totally repellent—in fact, he rejected quantum mechanics, the other great theory of modern physics, on exactly these grounds, finding the probabilistic nature of the theory too philosophically disturbing.*

In reality, relativity is a rigorous scientific theory, placing strict limits on what can be seen by different observers. Relativity gives different roles to concepts such as length and time that classical physics holds as absolute and unchanging, but it provides one, and only one, way of reconciling different observations. If you know what one observer sees, relativity lets you say with absolute certainty what any other observer will see.

Moreover, while relativity seems to upset the simple and orderly world of classical physics, looked at in the right way, it introduces a new and elegant symmetry into the description of the universe. When we look at space and time as different aspects of the same thing, we get a deeper insight into the way the universe works, one that becomes especially important as we move from the special theory of relativity, which deals only with observers moving at constant speed, into the general theory, which includes the effects of acceleration and gravity and provides the modern framework for understanding the history and eventual fate of the universe.

In this chapter, we look at the things that don't change in special relativity and use the new symmetry between space and time to introduce a graphical depiction of relativistic motion that makes some problems easier to understand. First, though, we need to talk about the classical conception of space and time and how physicists handle them mathematically.

---

* His feelings are summed up in the paraphrased statement "God does not play dice with the universe." Quantum mechanics and the philosophical problems that drove Einstein to reject the theory are explained for the interested canine reader in *How to Teach Physics to Your Dog* (New York: Scribner, 2009).

THE FUNDAMENTAL THINGS APPLY: SYMMETRY AND INVARIANT DISTANCES*

Back in Chapter 1, we talked about the way physicists deal with space mathematically, describing the position of objects in terms of their positions along three different directions: east-west, north-south, and up-down. The position of a squirrel relative to a dog would thus be described in terms of three numbers: 3m east, 4m south, and at ground level, say, or 1m west, 2m north, and 5m up a tree, annoyingly out of reach.

This representation is an effective tool for quantifying the behavior of moving objects, but it sometimes introduces confusion by making two situations that are really the same look different. The coordinate systems we use to describe the positions of objects are extremely useful but ultimately rather arbitrary. While there are some convenient reference points in the everyday world—for instance, the North Pole or sea level—the Earth does not come with preprinted lines of latitude and longitude. The choice of reference points is merely conventional, with humans agreeing to use the same system of coordinates to describe positions on Earth in order to simplify communication and trade between countries.**

The laws of physics do not depend on these coordinates in any fundamental way. All of the physical interactions that we consider behave exactly the same way in Schenectady, New York, as in Brisbane, Australia, or São Paolo, Brazil. When we think about physics, then, we should ask what quantities are the same no matter what coordinate system you use. These **invariant quantities** are what really matter in the physical world, and un-

---

* The section title is lifted from the song "As Time Goes By" from the movie *Casablanca*, as most humans and dogs would recognize. The full lyrics to the song include a little-known shout-out to relativity: "This day and age we're living in / Gives cause for apprehension / With speed and new invention / And things like the fourth dimension / Yet we get a trifle weary / With Mr. Einstein's theory / So we must get down to earth at times / Relax relieve the tension"

** The establishment of these conventions took decades of work from both scientists and diplomats, including Henri Poincaré, whom we met in Chapter 3.

derstanding how they behave gives us the most fundamental understanding of how the universe works.

The simplest example of an invariant quantity in everyday life is the separation between two objects. If you consider two objects—a dog and a bunny, say—their positions may be represented by very different sets of numbers using different systems of coordinates, but the distance between them will be the same. Emmy might measure the position of a bunny in her yard as 3m east and 4m north, while Zoe, the Weimaraner across the street, using a different system of coordinates would make it 3.54m in a direction a bit south of east and 3.54m in a direction a bit east of north. Both dogs would be 5m away from the bunny, though, and it's that distance that will ultimately determine the behavior of both dog and bunny.

In two dimensions, the east-west and north-south distances are two legs of a right triangle, so if we use x for the east-west separation between two objects and y for the north-south separation, we find the total distance d using the Pythagorean theorem:

$$d^2 = x^2 + y^2$$

A bunny that's 3m east and 4m north, then, is a total of

$$d = \sqrt{(3m)^2 + (4m)^2} = \sqrt{25}m = 5m$$

away from Emmy. According to Zoe, using a different coordinate system, the distance would be 3.54m in both the south-of-east and east-of-north directions, which is

$$d = \sqrt{(3.54m)^2 + (3.54m)^2} = \sqrt{25}m = 5m$$

or the same total distance. This Pythagorean rule will give you exactly the same distance between Emmy and the bunny, no matter how you orient your x and y coordinates.

When we include the third dimension of space, the up-down direction, we just extend this same rule, adding a third coordinate:

$$d^2 = x^2 + y^2 + z^2$$

A squirrel that is 4m east, 2m north, and 4m up a tree would thus be

$$d = \sqrt{(4m)^2 + (2m)^2 + (4m)^2} = \sqrt{36m} = 6m$$

away from Emmy on the ground barking angrily up at it.

"Wait a minute, dude. If there are three dimensions of space, shouldn't those be cubed, rather than squared?"

"No, squared is the right rule, no matter how many dimensions you have."

"But you used 'squared' for two dimensions—two dimensions, raised to the second power. So, three dimensions should go to the third power."

"You might think that, but you'd be wrong. The way to understand it is to realize that you can always orient your axes so that one of the components is zero. So, for example, if I used x to indicate the distance in a direction 30 degrees north of east and y to indicate 30 degrees west of north, then for the squirrel example, you'd have $x = 4.47$m, $y = 0$m, and $z = 5$m. In that case, you'd really only have two numbers to add, so you'd want the squared rule, not the cubed rule."

"Oh. That's sneaky."

"Maybe a tiny bit, but remember the important thing here: the distance between two objects is the same, no matter what coordinates you use to measure it. There will always be some arrangement of the x, y, and z directions that makes one of them equal to the distance and the other two equal to zero."

"So, everybody gets to choose her own coordinate system? Wouldn't it be easier to just get everybody to agree to use the same coordinates? Mine, for example, because I'm the best."

"It might be more convenient to have a single universal set of coordinates, but it doesn't matter. The things that are really important, in physics at least, don't depend on the coordinate system but on the distance between objects, which is the same for every observer."

"OK, I guess. But my coordinates are the right ones."

The fact that distance is invariant makes the choice of coordinate system unimportant in classical physics. The physical laws describing fundamental interactions between objects, such as the gravitational force or the electrostatic force, depend on the distance between them, not on the specific values of their coordinates in any system. The interactions of physics are symmetric, that is, the interactions between two particles follow the same rules when one particle is north of the other as when one particle is east of the other, or in any other direction you care to imagine.*

Most humans think of symmetry in the context of aesthetics, not mathematics, so it's natural to illustrate these concepts with pictures. When we think about a physical interaction—say, between Emmy in the middle of her yard and a nearby bunny—we can describe the position of the bunny relative to her using two perpendicular distances x and y.** We can make an accurate picture of the arrangement of Emmy and the bunny by putting the bunny x units to the right and y units toward the top of the page.

The choice of what directions to use as x and y, though, is ultimately arbitrary. So, while Emmy may say that x for a given bunny is 3m and y is 4m, Zoe, using a different definition of x and y, would record $x = -3.54$m and $y = +3.54$m. Winthrop the basset hound might use $x = 0$m and $y = -5$m, and so on. If we plot all the possible pairs of x and y values for the bunny, we find that they fall on a circle with a radius of 5m (see Figure 5.1).

The symmetry in the interaction between Emmy and the bunny means that all points on this circle are equivalent. You could move the bunny around to any point on the circle, and the result would always be the same: Emmy runs toward the bunny, and the bunny runs away. We could also reverse their roles, putting the bunny at the center and Emmy at points

---

* Some interactions, particularly magnetic interactions, also depend on the relative orientation of two objects—two magnets with their north poles together will be pushed apart, while two magnets arranged with the north pole of one facing the south pole of the other will be drawn together. The orientation is a property of the individual particles, though, and does not depend on the coordinates used to describe their positions.

** Technically, we need three numbers to describe the position in three dimensions. It's really difficult to represent a three-dimensional scene on a two-dimensional surface like a piece of paper, though, so we'll stick to earthbound creatures like dogs and bunnies to make it easier to draw the pictures.

on a circle around it without changing their interaction. The exact values of x and y for dog and bunny are unimportant; what really matters is the distance between them, which is the same for every point on that circle.

In classical physics, then, the symmetry of the dog-bunny interaction makes the invariant distance between Emmy and the bunny the essential quantity for understanding and predicting their behavior. When we formulate the rules governing the interactions between physical objects, we express them in terms of the invariant distance, not the specific values of x and y.

Figure 5.1.

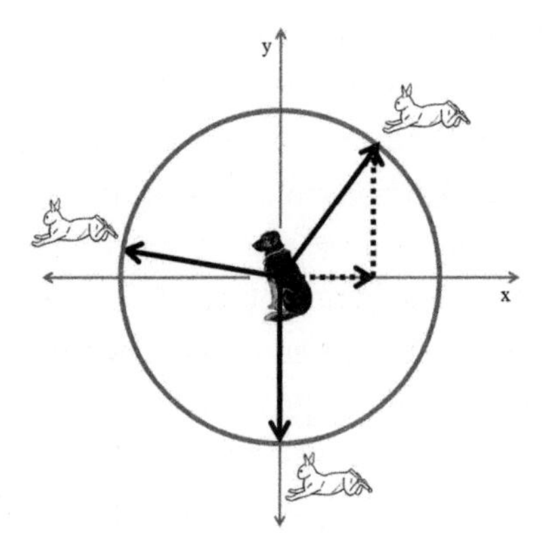

"You know, you're leaving out an important factor."

"Yeah, what's that?"

"Wind. It makes a huge difference whether I'm upwind or downwind of the bunny. The interaction can be completely different depending on who smells whom first."

"For the sake of the analogy, can we imagine that it's a calm day?"

"I guess so. Anyway, I have a bigger problem with this whole business."

"What's that?"

"Well, the coordinate system you're using is silly. I mean, why do you treat vertical distance the same as the other two?"

"Why wouldn't I? It's a dimension of space, the same as the other two."

"Yeah, but I can move east-west and north-south really easily, but I can't go up or down much at all. The up-down direction is completely different."

"So, you'd like to measure positions in terms of the distance from the house in some direction and the distance up and down?"

"Well, yeah. That makes so much more sense."

"Congratulations, you have rediscovered cylindrical coordinates. That's another way of specifying position using two distances and an angle. You keep track of the distance out from some central point, the angle away from some reference direction, and the height above or below some reference level. It's used when you're dealing with things that are cylinders—round in two dimensions and extended along a line in the third."

"So, how do you measure distances there?"

"It's the same method: you use the Pythagorean theorem with the distance out and the distance up. The angle just determines the direction, so it doesn't really come in."

"Oh. So, are there other systems too?"

"Sure, there are lots. The other popular one is spherical coordinates. There, you measure the distance from some central point and use two angles to specify the direction."

"Two angles?"

"Sure, like latitude and longitude. For example, we're a bit less than 43 degrees north of the equator and a hair under 74 degrees west of Greenwich, England. Those two angles specify our direction relative to the center of the Earth."

"Um, dude, that's only two numbers . . ."

"Right, because when we specify positions using latitude and longitude, we assume we're talking about a point on the surface of the Earth. So, to get to Niskayuna, New York, from the center of the Earth, you'd aim in a direction 43 degrees north of the equator and 74 degrees west of Greenwich, then travel about 6,400 km in that direction and end up here."

"That'd be hard to do, wouldn't it? I mean, that's a long way to dig . . ."

"Most of it's molten rock, too, so you wouldn't actually do it. But conceptually, that's what spherical coordinates are about."

"So, how do you decide which of these different systems for expressing position to use?"

"Well, you pick the one that makes the most sense given the geometry. If all you really care about is the distance away from some central point, you probably want spherical coordinates. If you've got something that moves according to one set of rules in two dimensions and a different set of rules in the third—like you running around the yard but being unable to fly or climb trees—then cylindrical coordinates might be the best choice. If you're looking at things that mostly move in straight lines, you probably want the Cartesian x, y, z system. You set it up so that your moving observer moves exactly along one of the directions, and the math becomes relatively simple."

"Pardon the pun."

"Yeah, pardon the pun. Anyway, while the choice of system may make some physical quantities easier to calculate, no matter what system of coordinates you use to designate a particular point, the distance between two points will always be exactly the same, no matter who measures it. That's an example of an invariant quantity, something that doesn't change depending on the observer."

"Except it does change in relativity—which is the whole problem."

"Right, but now that we've got the idea, we can talk about what *doesn't* change in relativity."

WORTH A THOUSAND EQUATIONS: SPACETIME DIAGRAMS

The ideas of symmetry and invariant quantities are incredibly powerful tools for theoretical physics. This was made especially clear in 1915 when German mathematician Emmy Noether* proved a mathematical theorem showing that symmetries in physics lead to **conserved quantities** like momentum and energy. Because the laws of physics do not depend on the

* No relation to Emmy the dog.

position of an object, **Noether's theorem** tells us that momentum is conserved, which provides a way to understand forces and collisions. Because the laws of physics are the same in the past, present, and future, Noether's theorem tells us that energy must be conserved, which provides a simple way of understanding interactions between objects in terms of energy flowing from one to another. Thanks to Noether's theorem, modern physicists look at the world in terms of inherent symmetries and the resulting invariant quantities. The symmetry in ordinary, everyday physics is more obvious than in relativity, but when we look carefully, we find symmetries and invariant quantities in relativity as well.

As we've already seen, the nature of space is very different in relativity: the distance between objects is different according to different observers. A stationary observer watching a moving object sees it shrink, and a moving observer sees the distance between two points in space as smaller than that measured by an observer who is stationary relative to those points. Distance in space is no longer an invariant quantity.

This might seem like a state of anarchy, with everything up for grabs, but last chapter's discussion of paradoxes points the way out of this potential mess: we can explain the change in an object's size measured by a moving observer by looking at the timing of the measurements. One observer measures the position of the ends of an object at a particular instant in time; a second observer moving relative to the first sees those two measurements separated by a smaller distance in space, but taking place at different times.

The key to finding an invariant quantity in relativity, then, is adding time into the mix. When we do that, we get back to a situation where all observers are in agreement, and we gain new insight into the way the universe is put together. The unification of space and time is one of the most profound consequences of relativity and one of the most powerful tools we have for understanding how the laws of physics work.

To explore this, we adapt the graphical technique of plotting the coordinates of an object in two dimensions by making time one of the two dimensions on the graph. We plot distance in the x direction along the horizontal axis and distance in time along the vertical axis.

"Wait a minute. What do you mean, 'distance in time'? Time's measured in seconds, not units of distance."

"Right, but we can make a time into a distance using the speed of light. Light moves at the same speed for all observers, so we can multiply the time by the speed of light and get a distance in time that is measured in meters, just like the distance in space."

"So, you measure time in meters?"

"That, or you could divide all the spatial distances by the speed of light and measure distance in seconds.* Either way works, but if you measure time via distance, that's only one dimension where the units seem odd."

"I guess one weird unit is better than three . . ."

This graphical representation of space and time is called a **spacetime diagram,** or Minkowski diagram, after the German mathematician Hermann Minkowski, who first recognized that relativity could be explained in geometric terms and was one of the most enthusiastic early adopters of the theory.** Minkowski introduced these diagrams in 1908, and they quickly became a standard tool for illustrating the effects of relativity.

In spacetime diagrams, the horizontal distance between points represents the spatial separation between two objects or events. The vertical distance between points represents the separation in time, so moving up along the diagram represents the passage of time into the future. Figure 5.2 shows an example of a spacetime diagram for the scene outside Emmy's window.

Every object in the diagram has a **worldline** representing its trajectory into the future. A stationary object like Emmy in her own frame or a tree across the street is represented by a straight vertical line: the object does

* This is done in astronomy, where distances are measured in light-years, the distance traveled by light in one year (9.5×10^{12} km).

** Minkowski famously asserted, "Henceforth space by itself, and time by itself, are doomed to fade away into mere shadows, and only a kind of union of the two will preserve an independent reality." This hasn't exactly panned out in the general culture, or even all of physics—there are lots of subfields of physics where it's perfectly valid to consider space and time separately—but at the smallest and largest scales of physics, it's absolutely true.

Figure 5.2.

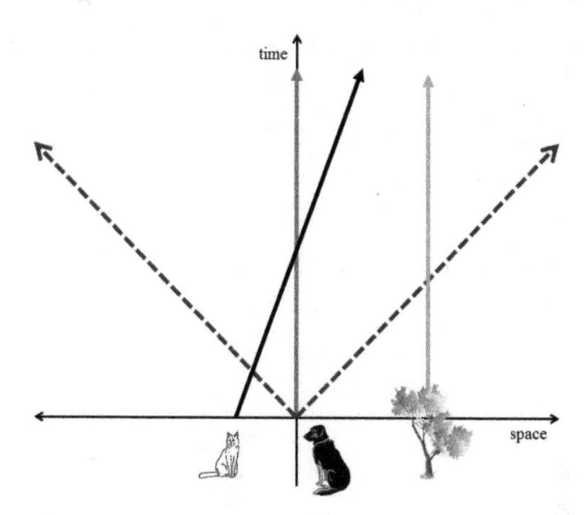

not move in space but marches relentlessly forward in time. A moving object, like that pesky Nero creeping around the neighborhood, shows up as a diagonal line, moving both upward into the future and sideways through space.

The diagram also includes two dashed lines at 45 degrees from the vertical, emanating from Emmy at the center. These lines are the worldlines for two rays of light leaving Emmy at the start of the scene, one headed left and the other right. These lines set the scale for the diagram: since we're using the speed of light to convert time to distance, light will always move one unit of distance in one unit of time, making a 45-degree angle. The speed of light is the same for all observers, so worldlines describing light rays will always be represented by 45-degree lines, no matter who's making the observations.

Each observer seeing a set of events in spacetime has his or her own diagram with him- or herself at the center. Figure 5.3 shows the diagram corresponding to the scene in Figure 5.2 from the cat's point of view.

According to Nero, he is stationary at the center of the universe, while other objects move around him. Thus, from Nero's point of view, both Emmy and the tree are moving, as represented by parallel diagonal lines slanting to the left. Again, we include two dashed worldlines for the two

Figure 5.3.

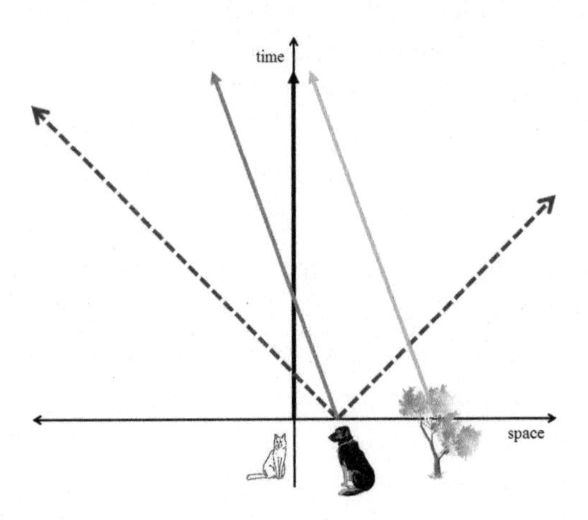

rays of light leaving Emmy at the start of the diagram; as in Emmy's diagram, these are at 45 degrees from the vertical because the speed of light is the same for both dog and cat.

Spacetime diagrams are useful for thinking about relativity, and the rules for converting from one observer's results to another's can be framed as rules for manipulating the positions of points and lines on diagrams of this sort. Even before we get to that, though, we can learn something important about the structure of spacetime just from thinking about these diagrams: we can see how to put together the invariant spacetime interval.

"What's to see? I mean, isn't it obvious?"

"What do you mean?"

"I mean, if time is just another dimension like space, there's nothing to see. You square it, and use the Pythagorean thingy to get the distance. End of story."

"You might think that, but you'd be wrong."

"But you said time was just another dimension, and when we added more dimensions of space, you added them Pythagorean-like."

"I said that time and space were different aspects of the same thing, not that they were identical. They're obviously different, because you can only move forward through time, not backwards. That difference leads to a different form of the spacetime interval."

"How? And how do these diagrams help?"

"Watch and see."

So, what is the appropriate rule for constructing a spacetime interval? We can find it by looking at how two events—say, Emmy waking from a nap, then barking at something a short time later—appear to different observers. By plotting these two events as they appear to different observers on a spacetime diagram, we can find the pattern followed by these observations and trace out the symmetric curve of points having the same spacetime interval.

From Emmy's point of view, of course, both her waking and barking take place at the same position, at two different times. The point corresponding to her barking is thus directly above her waking on the spacetime diagram. Nero moving by from left to right, on the other hand, who happens to be at the same position* as Emmy at the time when she wakes, sees the barking event taking place at a position to the left of the center of his diagram, because he's left Emmy behind by the time she starts barking. Nero will also see the barking event taking place at a later time than Emmy, so the point will be higher up (see Figure 5.4).

"Wait a minute—the cat sees things take *more* time? I thought moving clocks ran slow?"

"They do. In Nero's frame of reference, though, it's *your* clock that's moving and thus running slow. According to him, you see a shorter time between waking and barking because your clock is ticking at a slower rate than his. Therefore, Nero sees more time pass between waking and barking."

---

* The same east-west position, that is—they're obviously not at the same position in all three dimensions, because that would lead to major trouble. We only plot one dimension of space in the diagram, because we're stuck using two-dimensional paper.

"Oh, OK. Hey, is that why the Pythagorean rule doesn't work?"

"You're absolutely right. If we just added the time distance according to the simple Pythagorean rule, the observations would fall on a circle around the waking up, like with ordinary distance. That would require *less* time to pass according to a moving observer, to make up for the increased distance in space. But that goes against what we already know about the relativity of time. Good catch."

"I'm a highly observant dog, dude. You can't get anything past me."

If we look at a third observer's view of events—say, a bunny who gets spooked when Emmy wakes up and who runs in the opposite direction from Nero at a higher speed—the barking takes place even later in time than in Nero's observation and farther away from the center of the diagram in the opposite direction. This point is above and to the right of Emmy.

If we repeat this process many times, plotting the spacetime location of Emmy's bark according to observers moving at different speeds and in different directions on a spacetime diagram, we find that all the points fall on the black curve in Figure 5.4. As the speed of the observer relative to Emmy increases, her barking moves farther away in space and later in time. The resulting curve is one well known to physicists and mathematicians: a hyperbola.

Figure 5.4.

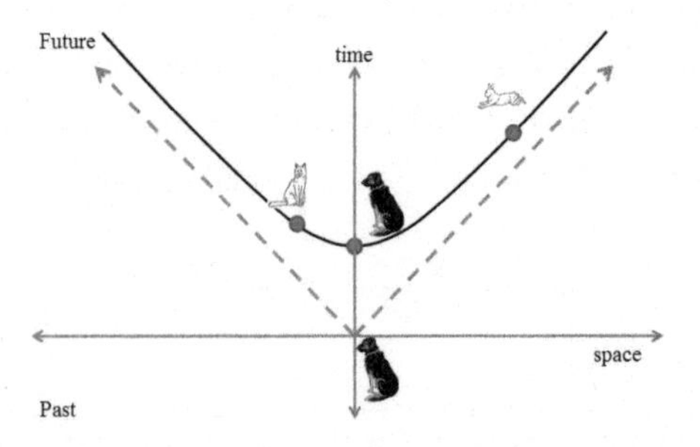

"Hah! I knew it! You're just making this all up!"

"What? What are you talking about?"

"Dude, you just admitted it yourself. This whole business is a wild exaggeration. A hyperbola."

"A wild exaggeration is a hyperbole, not a hyperbola. A hyperbola is a mathematical curve having a particular shape. It turns up all over physics, in things like the path followed by a comet falling into the solar system or that followed by a particle subjected to a repulsive force in a collision. They're completely different words."

"Oh. Well, hyperbole, hyperbola. Let's call the whole thing off."

"You're ridiculous."

The hyperbola is a well-known mathematical function, first studied by the ancient Greeks. It tells us that the correct definition of the invariant spacetime interval Δs is*

$$\Delta s^2 = x^2 - (ct)^2$$

This equation summarizes everything we have discussed about relativity. As the spatial distance between two events increases, the distance in time must also increase in order to keep Δs the same. Once we locate an event as a point on a spacetime diagram, we know that the same event according to other observers will always fall on a hyperbola that passes through that point. The rules for converting from one observer to another are thus geometric rules for moving points along these hyperbolic curves. Hermann Minkowski was the first to explain relativity in these geometric terms, an insight that points the way toward the general theory of relativity, Einstein's greatest triumph, which explains gravity as a warping of the geometry of spacetime (Chapters 9 and 10).

"That's cool and all, but there's a problem with your theory."

---

* There are two ways of writing the spacetime interval: you can subtract distance in time squared from the distance in space squared, as above, or you can subtract the distance in space squared from the distance in time squared. Inevitably, two different communities of physicists each favor a different approach: roughly speaking, physicists who primarily work with general relativity tend to use the time-subtracting version used above, while particle physicists use the distance-subtracting version.

"Which is? . . ."

"Well, looking at your equation there, you've got Δs squared equal to the difference between the time and space differences."

"Yeah."

"Well, that can't be right, because if the distance in time is greater than the distance in space, then the spacetime interval squared is negative. And you can't take the square root of a negative number. Not even with a calculator. I've tried."

"It's not actually a problem. Mathematicians and physicists deal with the square roots of negative numbers all the time—they're called imaginary numbers and are incredibly important for physics, particularly quantum mechanics."

"So, the spacetime interval is imaginary? This whole relativity business is a figment of your imagination?"

"No, it's only sometimes imaginary. Other times it's a real number, depending on whether the spatial distance is bigger than the time distance, or vice versa. That's a good catch, though, because the different signs of Δs^2 divide spacetime up into regions. It's a great lead-in to the idea of the **light cone**."

"Oooh! I love those! They're all crunchy and sweet and cold and yummy . . ."

"What?"

"We are talking about low-fat ice cream, right?"

"No, we're not. We're talking about physics. Try to concentrate, please."

"I'll try, as long as we can go out for ice cream later."

YOU CAN'T GET HERE FROM THERE: THE LIGHT CONE

The definition of the spacetime interval above makes geometric sense, but if you're quick with arithmetic, you may notice an apparent problem: when the distance in time between two events is greater than their separation in space, as it is for the example events we've been talking about, the formula gives a negative value for the spacetime interval squared. You may even have noticed that you can imagine another hyperbola with the

Figure 5.5.

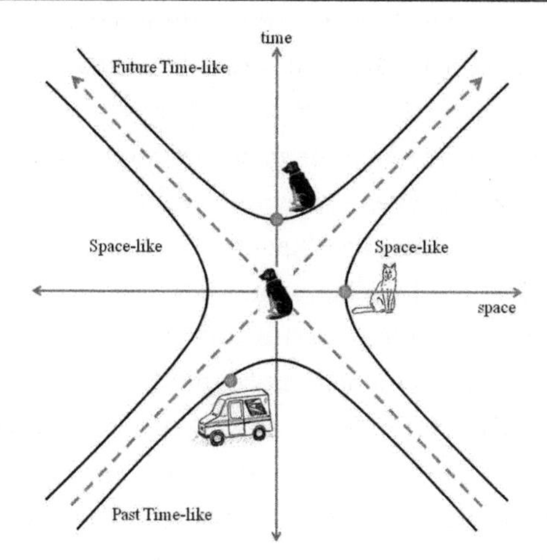

same magnitude of the spacetime interval squared (that is, with a positive value the same as the negative value for the original hyperbola), where the separation in space is greater than the distance in time. Rather than being a flaw in the definition, though, this allows for an extremely useful categorization of events.

In fact, the hyperbola we drew above is one of four curves having the same magnitude of the spacetime interval squared, two with positive values and two with negative values. Figure 5.5 shows the full graph.

Looking at these curves, one important feature jumps out: each of these four hyperbolae is confined to one of the four regions created by the 45-degree lines representing light pulses sent out from (or arriving at) the moment when Emmy wakes from her nap. While each of the hyperbolae crosses one of the axes of the graph, none of them cross both.

These facts let us divide the spacetime diagram into four regions depending on the type of separation between the events in those regions. The top and bottom hyperbolae, for which the spacetime interval squared is negative, have **time-like separation** from the central event of the dog waking up. The top hyperbola represents all the possible observations of an event in Emmy's future, like her barking at some interloper. While there are some observers who will see this event happen at the same place as the waking (including Emmy herself), no observer will see the two events happen at the same time. The bottom hyperbola represents all the possible observations of some event in Emmy's past—such as the mail carrier stepping onto the porch. Again, while some observers will see this event and the waking happen at the same place, no observer will see the two events happen at the same time.

The other category of events, represented by the hyperbolae on the left and right, for which the spacetime interval squared is positive, have **space-like separation.** The hyperbolae on the left and right represent events that take place some distance away from Emmy waking up—a neighbor's cat scratching at the furniture, say—and while there are observers who will see these events take place at the same time, no observer will see them take place at the same position.

"Seriously, dude? The best you could come up with is 'time-like' and 'space-like'?"

"What?"

"You physicists really do stink at names, don't you?"

"Do you have a better suggestion?"

"Ummmm . . . no, not at the moment. But then, Human isn't my first language."

An important feature of the graph in Figure 5.5 is that none of the hyperbolae ever cross the 45-degree line representing a beam of light sent out from the first event. As the speed of the observer gets faster and faster, the events will come closer to that line but will never reach it, let alone cross it.*

The fact that the light beams neatly divide these curves gives a name to the categorization of the curves: points inside the regions of time-like separation are within the "light cone" of the initial event. The name comes

* This is another indication that the speed of light is a fundamental limit, which we will explore more in Chapter 6.

because if you add a second dimension of space (in and out of the page) to the graph, the *V* shape of the two 45-degree lines will rotate into a shape like two cones meeting at their points.* In reality, of course, there are three dimensions of space, so the cones are really four-dimensional figures representing a sphere expanding in three dimensions at the speed of light. Until somebody invents four-dimensional paper, though, we'll stick with "light cone" as the name.

The light cone has great importance in relativity because points within the light cone of an event can be connected to it by some chain of cause and effect. That is, an event in Emmy's past light cone, such as the mail carrier stepping onto the porch, can be the proximate cause of her waking up. Waking up is followed by looking outside, which leads to barking. The barking then causes the mail carrier to retreat from the porch,** and so on. All of the events in this chain of cause and effect fall within the past and future light cones of Emmy's waking.

Events falling outside the light cones, with space-like separation from the central event, can never be the cause of, or be caused by, the event at the center of the light cones. Any observer looking at these events sees them falling into one of the hyperbolae to the left or right in Figure 5.5. This means that different observers may disagree about the order in which the central event and a space-like-separated event take place—one sees it happen before Emmy wakes and the other afterwards—but this is not a problem, as the two cannot be causally related. Moving a space-like-separated event from past to future will never result in an effect appearing to happen before its cause, so some measure of sanity is preserved.

An event with space-like separation from Emmy's waking can be the cause of some event in her future light cone—the cat scratching the furniture may lead to the cat being thrown outside, which will then trigger more barking—but not until enough time has passed for the scratching

---

* In this three-dimensional version, the left and right hyperbolae rotate into one another, forming a single, space-like separated surface shaped sort of like the neck of a vase humans might use to put flowers in.

** Humans might object that the carrier is just following a predetermined route and never intended to stay on the porch for more than a few seconds, but dogs know the truth.

cat to fall within the past light cone of the second barking episode. Similarly, an event with space-like separation from the original event can also be caused by some event in Emmy's past light cone—the arrival of the mail may also have startled the cat into scratching the furniture—but she won't know about it until enough time has passed for it to come within her past light cone.

The light cone for a particular observer is not a fixed and unchanging thing but moves with the observer as he or she advances through time. You can imagine your own past and future light cones like two funnels sliding along a wire representing your worldline. As time goes on, the space encompassed by your past light cone expands, eventually including the entire universe. The critical limit is the speed of light: one event can cause or be caused by another event only if the two are separated by enough time to allow light to travel between them.

The spacetime interval defined above provides not only an island of stability in the sea of changes predicted by relativity but a neat way to categorize events in spacetime. We can only affect (or be affected by) events that fall within our light cones, and cause will always precede effect for every observer, no matter how fast he or she is moving.

"So, basically, the light cone is like one of those restrictor cones vets put on dogs after surgery?"

"Hmm?"

"You know, those big plastic things that wrap around your neck, so you can't lick the itchy stitches. Events that fall within the cone—say, a treat falling into it—are time-like separated, because it's only a matter of time before you get the treat and eat it. Things that fall outside the cone—like the stitches or bandages—are space-like separated, because you'll never be able to reach them, no matter how badly you want to."

"Yeah, that's a pretty good way of putting the restrictions of the light cone into canine terms. Good job."

"Thank you. I am the best, you know. One thing, though."

"What's that?"

"I hate those stupid cones. All dogs do."

"True enough. Of course, they're good for you in that they help you heal properly, in the same way that the light cone in relativity is good for you in that it ensures that **causality** is always preserved."

"I guess so. Still, regarding those cones: *do not want!*"

"So noted."

SEEING BOTH SIDES: MULTIPLE OBSERVERS ON A SINGLE DIAGRAM

To this point, we have discussed the behavior of space and time and the use of spacetime diagrams with the help of a generic diagram that represents measurements by different observers as points on a single diagram. We can also use a diagram combining the worldlines of multiple observers to show what each observer would see looking at the same set of events. This technique provides a convenient pictorial method for illustrating how different observers view the world and can shed some additional light on the central issues of clock synchronization and time dilation.

To see how to present the measurements of two different observers on a single diagram, it's helpful to look at two diagrams representing a very simple set of events: Emmy sitting in her living room watches Nero move past at half the speed of light (see Figure 5.6).

Figure 5.6.

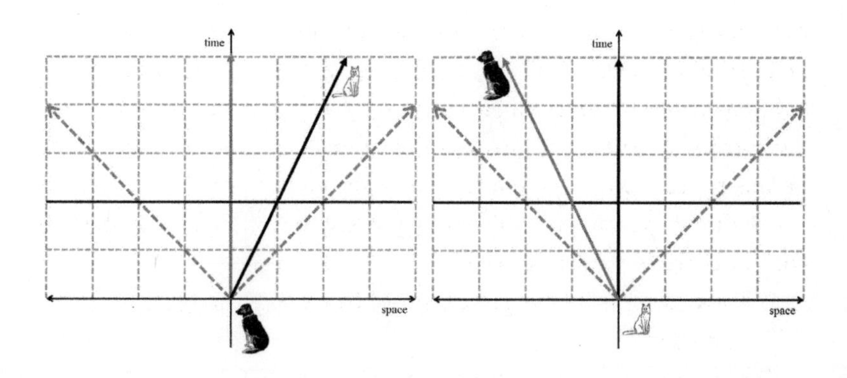

To simplify matters, we start both diagrams at the instant when Nero and Emmy have the same east-west position. On the left, we see the diagram from Emmy's point of view, in which she is at rest at her starting position, and on the right we see the diagram drawn from Nero's point of view. Both diagrams also include two lines representing light emitted at the instant Nero and Emmy are at the same position. Since the speed of light is the same for both, the light rays make a 45-degree angle in both diagrams, but other than that, we can't immediately say what specific point in Nero's diagram corresponds to a given point in Emmy's.

So, how do we obtain Nero's measurements by looking at Emmy's diagram, or vice versa? This may seem an impossible task, but in fact, Nero's space measurements are easy to place on the diagram. Any cat always measures position relative to himself, which means that whatever markers Nero uses to define the position of objects in his frame must have worldlines parallel to his. Thus, we can draw a partial grid on the diagram representing Nero's position markers by dot-dot-dash lines parallel to his worldline. Each line represents one unit of distance away from his position (see Figure 5.7).

Figure 5.7.

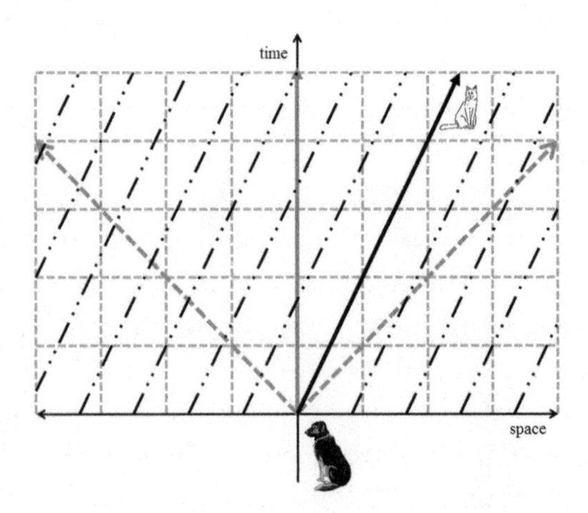

"Wait a minute. Those one-unit lines aren't the same distance apart as my one-unit lines!"

"That's right, because Nero is moving relative to you. You see his length measurements as shorter due to length contraction."

"Yeah, but they don't even line up with my grid at all."

"Again, that's because Nero's moving. The point that he sees as being one unit from himself must move with him—which means that all his markers change position as time goes by, so they're not parallel to your position lines."

"Oh, OK. So, what about the grid for his time measurements? They must slope down and to the right, right?"

"Wrong. Actually, they slant up and to the right."

"What? That doesn't make any sense! That means his grid won't be at right angles."

"It makes reading Nero's measurements off your diagram a little more complicated, to be sure, but that's the right answer. It has to be, and you can see it from the diagrams. Watch."

To complete the process of locating Nero's measurements on Emmy's diagram, we next need to consider how to place a grid of lines representing his measurements of time on the diagram. The key to drawing the time grid is the speed of light. We know that both Emmy and Nero always see the outgoing light rays at any instant (represented by the horizontal line in each diagram) at the same distance away from their position, because the speed of light is the same for both. Looking at the positions of the animals relative to the light pulses in the paired diagrams in Figure 5.6, we see that Emmy always sees the distance in space between herself and Nero as equal to the distance in space between Nero and the rightward light ray. At the particular instant in time represented by the dark horizontal line—two time units after the start—he is one unit to her right, and the light is one unit to his right. Meanwhile two units after the start of his diagram, Nero sees the same distance in space between himself and Emmy as between her and the leftward light ray, one unit each.

To put Nero's measurements on Emmy's diagram, we need to identify points where the distances from cat to dog and cat to light ray have the proper proportion, using our slanting grid of position lines. Two units after the start of the diagram according to Nero, the light rays should be two units away from him in either direction, and Emmy should be one unit to his left. If we draw a line that crosses the worldline for the leftward light ray when it is two cat units to his left, Emmy's worldline where it is one cat unit left, and the rightward light ray when it is two cat units right, we get the dark dot-dash line in Figure 5.8.

Figure 5.8.

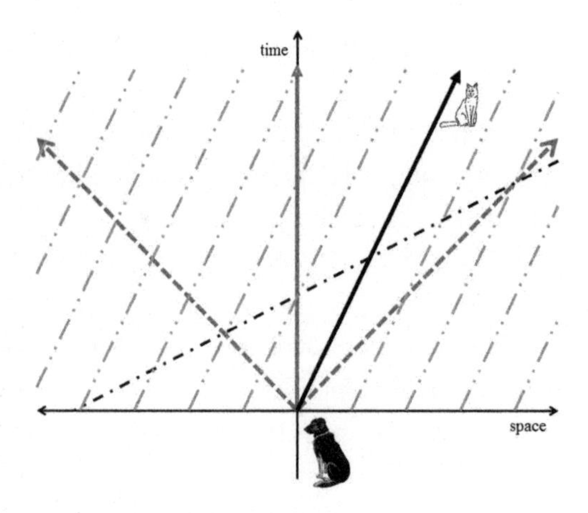

This sloping line represents a single instant in time in Nero's frame, the instant two cat units of time after the start of the diagram. According to Nero, all the events whose points in Emmy's spacetime diagram fall on the dot-dash line happened at precisely the same time. Events taking place later in his frame of reference will fall on a line parallel to this one, but higher up, and so on. We can thus cover the diagram with a grid of slanted lines representing steps in time and space according to Nero and use this grid to measure the position and time he sees for events graphed in Emmy's frame of reference (see Figure 5.9).

Figure 5.9.

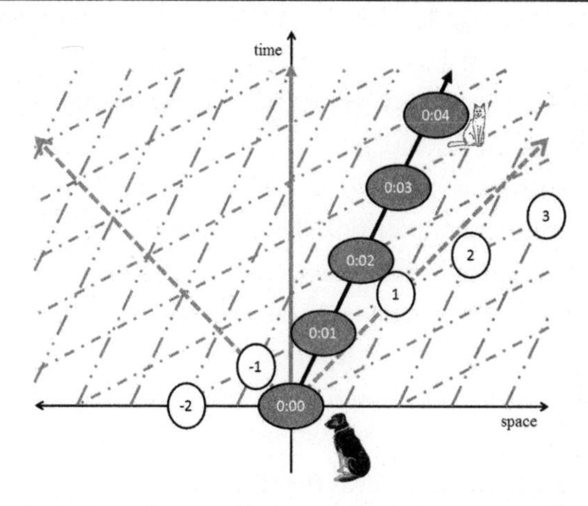

As time moves forward in Nero's frame, he moves up and to the right in Emmy's diagram. Particular instants in time are marked by the dot-dash lines sloping up and to the right, labeled by the clocks along Nero's worldline. The position of objects in space is measured against the grid of dot-dot-dash lines, parallel to Nero's worldline, and labeled at the instant one unit of time after the start of the diagram. Notice that for both Nero and Emmy, the two rays of light each move one unit of space for each unit of time, so they both see the same speed of light.

Of course, as with everything in relativity, the process of making these diagrams is symmetric. Nero could equally well choose to plot all of Emmy's measurements on his own diagram, in which his measurements of time and space would fall on horizontal and vertical lines, while hers fell along a grid that slanted to the left (see Figure 5.10).

All of the observations we made about Nero's measurements in Emmy's diagram apply equally well to her measurements in his frame. As this is a book about explaining relativity for dogs, we'll mostly stick to diagrams drawn from the dog's point of view, but it's important to know that they can be drawn the other way as well.

"So, wait, how does the cat know about all these distant points, anyway? I mean, how does this fit with all that light cone business?"

"What do you mean?"

"Well, you're establishing an instant of time by recording the position of two light rays leaving the cat. But those measurements happen far from the cat, so how does he know where they are? Shouldn't they have space-like separation from the cat?"

"Ah, yes. That's a good catch. The measurements of the light ray positions at any given instant do indeed have space-like separation from Nero at that instant. He can't make those measurements himself—he needs helper cats at different positions making conscientious measurements of when the light rays cross their positions."

"But how does he get the results of those other cats' measurements instantaneously?"

"He doesn't. Each cat records the time of events at his own position, and they all get together later to reconstruct what happened at each position at various times. When they do that, they can map out the worldlines of the light rays and other objects and use that to put together the diagram."

"Yeah, well, good luck with that."

"With what?"

"With getting a bunch of stupid cats to agree about a diagram. People joke about 'herding cats' for a reason, you know."

Figure 5.10.

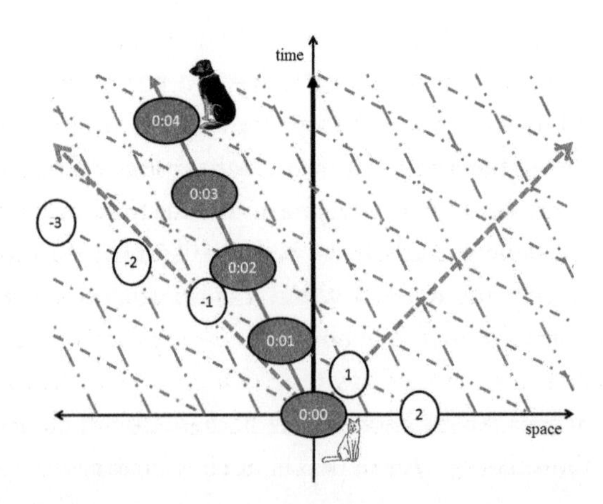

These spacetime diagrams provide a nice representation of the problem of synchronizing clocks discussed in Chapter 4. If we imagine Emmy conscientiously preparing a huge array of synchronized clocks at different positions, for use in recording events in spacetime, we get a picture like the left side of Figure 5.11, with all the clocks along the same horizontal line showing the same time, corresponding to a particular instant in her frame. To Nero, though, who sees events at the same time falling on the slanted lines, these clocks are not properly synchronized. At the instant represented by the shaded region in the diagram on the left, the clocks to the right are ahead of the clocks to the left, and the farther left or right you go, the greater the disagreement.

Figure 5.11.

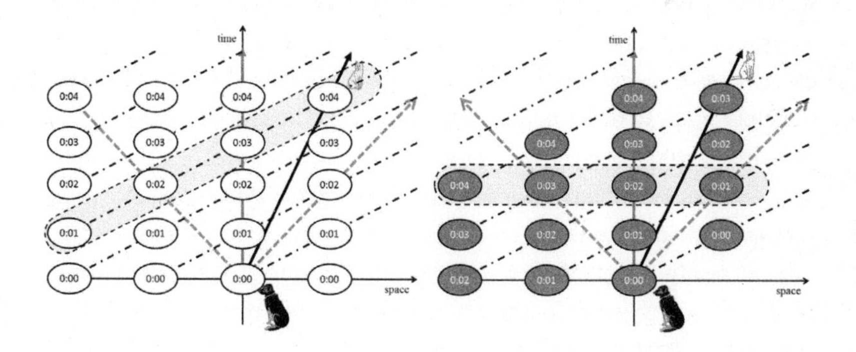

The synchronization problem is symmetric. If Nero prepares a set of synchronized clocks in his frame, we get the situation shown on the right side of Figure 5.11. According to Emmy, at the instant represented by the shaded horizontal region in the diagram, Nero's clocks are out of synch with one another. She sees clocks to the left as set ahead of clocks to the right in exactly the same way that her clocks look out of synch to him.

These diagrams also let us see how both Emmy and Nero can feel that the other's clock is running slow, as seen in Figure 5.12. According to Emmy, at the instant when twelve units of time have passed on her clock, Nero's clock shows only ten units, so his clock is running slow. According

Figure 5.12.

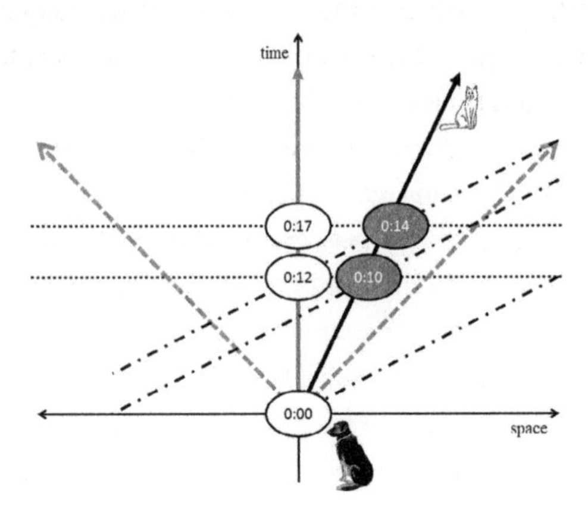

to Nero, though, the instant when Emmy's clock shows twelve is really represented by a point on a slanting line, crossing his worldline when his clock reads fourteen, so he says her clock is running slow. Emmy thinks her clock is ahead of his because she's looking at his clock in the past. And, of course, from Emmy's point of view, Nero is looking at her clock in the past—when his clock reads fourteen, hers reads seventeen.

Spacetime diagrams can be used in this manner to explain most of the strange results of relativity. They take some getting used to but provide a powerful tool for understanding how motion affects time and space. We'll use them many times in the following chapters.

"OK, I have a big problem with this."

"What's that?"

"Well, I like that you're representing everything from the dog's point of view, as is right and proper, but I don't like the way you've represented the cat's clock in this last picture."

"What's wrong?"

"Well, you've got my clock covering twelve units of time as it moves straight up. But the cat's clock only covers ten units of time as it moves up and to the right."

"Yes, and? . . ."

"But his path is longer than mine! How can it take less time to move a longer distance? I mean, that's what set us off on the whole time dilation thing with the triangle clocks."

"Ah, yes. That's an important difference between the pictures showing the light clock and these spacetime diagrams. Spacetime diagrams are an abstract representation, not a perfect match for ordinary reality. A longer path on the diagram corresponds to a shorter time for the clock following that path."

"Yeah, but how is that?"

"Well, a path that covers a greater distance on the diagram moves more in space than the shorter path, right?"

"Right."

"Which means that it's moving faster than the object taking the shorter path. And moving clocks run slow. So they must see less time pass than clocks that are at rest or moving more slowly."

"I guess, but it's still kind of weird."

"Yes, it is. The geometry of spacetime isn't the same as the geometry of ordinary space, so you need to be careful in reading these diagrams. Getting quantitative results from them—exactly how long a moving object appears or exactly how much time passes according to a moving observer—takes a lot of practice. They're really good for seeing qualitative results, though, like what order different observers see things happen in. And they're great for determining what's in the light cone of a given event and what isn't."

"I guess. So, hey, a faster-moving observer would have a worldline whose slope is closer to 45 degrees, right?"

"Yes."

"And the lines representing an instant in time squeeze up toward the 45-degree line, too, right?"

"That's right."

"So, what happens when you move faster than light? Do the space and time axes switch places or something?"

"No, because you can never go faster than the speed of light."

"Why not?"

"For a whole bunch of reasons, which we'll talk about in the next chapter, OK?"

"OK, but can we get ice-cream light cones, first?"

Chapter 6

299,792,458 M/S ISN'T JUST A GOOD IDEA, IT'S THE LAW: VELOCITY, MOMENTUM, FORCE, AND THE SPEED OF LIGHT

I'M IN THE LIVING ROOM grading student homework with the Giants game on TV. The game goes to a time-out, and a loud used-car commercial starts up. As I grab for the remote, I bump into a glass on the table, slopping water all over a stack of papers. I curse and mute the TV while I head to the kitchen for something to mop up the spill.

Emmy, who until this point has been curled up next to the couch, explodes into motion, running in a big figure-eight loop through the living room, into the dining room, around the table, and back to the living room.

I watch her complete a couple of orbits, then ask, "What in the world are you doing?"

She skids to a stop, panting. "I'm trying to help!"

"How does careening around the house help with the water spilled on my papers?"

"I'm just building up speed. When I get going fast enough, I'll go backwards in time and fix it so you don't spill the water in the first place."

"Uh-huh. Were they playing *Star Trek IV* on cable again?"

"I don't think so. I hope not. I'd hate to miss it." She's a big fan. "I'm just making relativity work for me." She wags her tail really proudly.

"Yeah, about that . . ."

"I figure that, since moving clocks run at different speeds, if I can move fast enough, the clock will run backwards, and I can go back in time!"

"I hate to burst your bubble, but relativity doesn't work like that."

"Why not? Moving clocks run slow, and the faster they move, the slower they run, so there must be a point where they run so slow, they run backwards."

"Time dilation gets bigger the faster you go, true, but it never goes backwards. As you get closer to the speed of light, a moving clock almost stops, but it never completely stops, and it certainly doesn't run backwards."

"Yeah, but what about moving faster than light, hmm?" She wags her tail like she's really got me. "I bet they go backwards if you go faster than light."

"It really doesn't matter. No material object can move faster than light, so you're never going to find out."

"What do you mean?"

"I mean that one consequence of relativity is that nothing with mass can ever reach light speed. You can get really close, but never attain the exact speed of light, let alone move faster than light."

"Why not?"

"Well, one reason is the thing that you touched on. If you could send messages faster than the speed of light, you could set it up so some observers would see effects coming before the events that caused them, which would be really weird."

"Yeah, but 'it would be really weird' isn't much of a scientific principle. I mean, physics involves a lot of really weird stuff, and you're fine with that. What makes cause-and-effect special?"

"Another reason is that anything one observer sees as being slower than the speed of light will always be slower than the speed of light, no matter who looks at it."

"Wait, what?"

"Yep. If you look at the way velocities add in relativity, you can never get anything moving faster than the speed of light."

"Sure you can. You just get a rocket ship moving at 99.9 percent of the speed of light and have it shoot out a second rocket at 99.9 percent of the speed of light. Somebody watching that go by would see it moving faster than light, wouldn't they?"

"No, they wouldn't. They'd see it moving at 99.99995 percent of the speed of light."

"OK, but that's only short by what, 100 m/s?"

"More like 150 m/s, but go ahead."

"So, you just give it a little more push, you know, to get it over the hump." She wags her tail hopefully.

"And there's the most practical reason: the force required to accelerate an object gets bigger as you get closer to the speed of light. It would take an infinitely large force to accelerate an object to exactly the speed of light in a finite time. And it would take infinitely long for any finite force to accelerate an object to exactly the speed of light."

"Really?"

"It's true. If you want to increase the speed of a 1 kg object by 1 m/s in one second, you need a force of 1 newton (N) if it starts from rest, right?"

She gives me a pitying look. "Thank you for reminding me of the definition of 1N."

"Well, if that object moves at 99.9 percent of the speed of light, you'd need 22N to get the same acceleration. And at 99.99995 percent of the speed of light, you'd need almost 1,000N to get the same acceleration."

"Oh."

"Yeah," I scratch behind her ears. "You can pull a leash with the best of them, but nowhere near that hard."

"OK, so *I* can't get going faster than the speed of light from rest, but what about someone who started out faster than light? Couldn't they, you know, pick me up and carry me along?"

"Someone starting faster than light would have the opposite problem—they wouldn't be able to slow down to the speed of light. But that's not important, because faster-than-light beings don't exist."

"They don't?"

"Lots of exotic physics theories end up predicting faster-than-light particles, called *tachyons*, but they're usually a sign that somebody made a mistake. Even if they could exist, they couldn't interact with ordinary matter, because if they did, it would make a mess of causality. There are people out there who study the properties that these particles would have, but it's a fringe thing. Most physicists are pretty certain that they don't exist."*

"Oh."

"So, no faster-than-light travel for you. Sorry."

"Don't apologize to *me*."

"Why not?"

"I was just trying to help. If I can't go backwards in time, you're the one who has to mop up the spilled water." I had forgotten the papers while talking about physics and groan when I look back over at the sodden mess. "That looks like it might take a while, dude," she says. "As long as you're up, how about letting me outside?"

One of the best-known consequences of relativity is that nothing can travel faster than, or even reach, the speed of light. This is a tremendous disap-

---

* In September 2011, an experiment in Italy released a preliminary result showing that the exotic particles known as neutrinos seem to move at 1.000025 +/− 0.000004 times the speed of light, in contradiction to relativity. While this would be a revolutionary discovery if true, most physicists—including some of the experimenters who measured it—suspect that it stems from some subtle error in the experiment and does not really prove that neutrinos are tachyons. Another experiment, at Fermilab in Illinois, is making a similar measurement that could confirm or reject the Italian result, which should be completed around the time this book is published.

pointment to dogs who want to chase alien bunnies on planets orbiting distant stars, as it means that reaching the nearest star would take at least four years (according to an observer remaining behind on Earth) with the most absurdly optimistic space technology and several decades under more realistic assumptions.

While science fiction writers have spent many hours imagining ways of getting around this problem (the "warp drive" of *Star Trek* and "hyperspace" of *Star Wars* being typical examples), this speed limit is one of the central results of relativity. In this chapter, we discuss three reasons why nothing can go faster than light: the relativistic rules for adding velocities, which show that anything moving slower than the speed of light according to one observer will always be seen moving slower than light by any other observer; the relativistic definition of momentum, which shows that an infinite force would be required to accelerate anything up to the speed of light; and finally, the fact that if faster-than-light travel were possible, it would violate causality.

1/2 + 1/2 = 4/5: RELATIVISTIC VELOCITY ADDITION

We've seen that relativity requires a rethinking of where and when events take place according to observers who are moving relative to one another. These changes force us to reconsider the velocities of moving objects seen by moving observers. We saw back in Chapter 1 that the velocity of an object is the change in its position over time, so when moving observers disagree about both the size of objects and the timing of events, they clearly must disagree about the speed of moving objects.

We can see how this plays out using spacetime diagrams. For simplicity, we'll consider the worldlines of three objects: Emmy in her usual spot at the window, Nero moving from left to right at half the speed of light, and a bunny moving in the opposite direction, also at half the speed of light. All three have the same position in space at some instant, which we'll put at the center of our diagram. Figure 6.1 shows the diagram from Emmy's point of view.

We want to know how fast the bunny is moving according to Nero (or vice versa). The Galilean rules from Chapter 1 would have us simply add the speeds, so Nero would see the bunny moving at exactly the speed of light. In relativity, however, the notion of time and space changes for a moving observer, so we have to replace the Galilean rules with something consistent with the principle of relativity.

We can approach this problem by thinking about how Emmy determines the bunny's speed from her own diagram. The velocity of the bunny depends on the distance it moves and the time taken to move that distance, which we measure by putting a grid over the spacetime diagrams. The bunny takes four units of time to move two units left; thus, its speed is half the speed of light, the same as Nero's speed moving to the right.

To get the bunny's speed seen by Nero from Emmy's diagram, we use the same process of overlaying a grid on the diagram. In this case, though, we need to use the tilted grid from the last chapter. Figure 6.2 shows Emmy's diagram with Nero's grid overlaid.

Following the worldline of the bunny from Emmy's position to the point marked on the diagram,* we see that it crosses four dot-dashed space lines and five dot-dot-dashed time lines. Thus, Nero says that the bunny's speed is four-fifths the speed of light.

"Your neat little boxes have become funny parallelograms."

"Technically, I think it's a rhombus, because all four sides are the same length, but yes, that's right."

"Isn't that kind of screwy?"

"Well, that's spacetime for you. What you see as distance only in space or only in time Nero sees as distance in both space and time."

"OK, but how do you know how to make the rhombuses?"

---

* This is not the same point as the point we used on the dog's grid. According to the dog, this point is a spatial distance of 1.73 units left and a time of 3.46 units after the start of the diagram. Any point on the bunny's worldline will give you the same speed; this point happens to have nice round numbers for its coordinates in the cat's frame.

Figure 6.1.

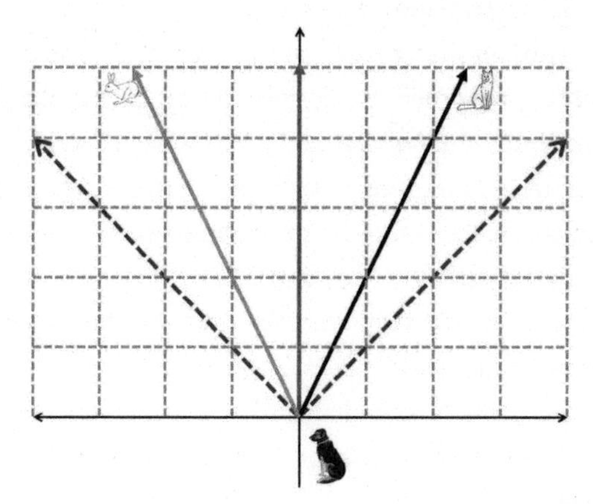

Figure 6.2.

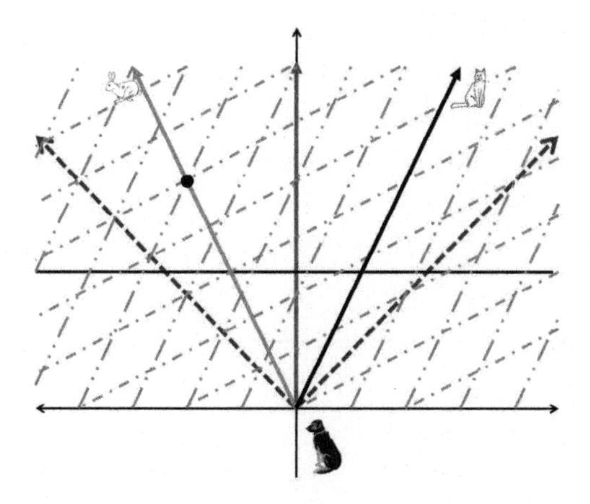

"In graphical terms, the rule is that the area of a single rhombus has to be the same as the area of the squares on your diagram. You can also just remember that moving clocks run slow by the Lorentz factor, γ, and use that to set the scale."

"It must be weird to be a cat and use those rhombuses as your grid lines instead of nice, sensible squares."

"No, it's not. At least, it's not weird to be a cat for that reason—in Nero's frame, the gridlines would be squares, and your gridlines would make rhombuses slanting the other way. The diagram drawn by Nero would look like that shown in Figure 6.3."

"Yeah, that's not all that weird."

"That's right. Remember, in relativity, everything looks perfectly reasonable to each individual observer. It's only when you try to compare measurements across different frames of reference that anything looks weird, as you can see from the slanting grid representing the dog's measurements on that figure."

"So, why didn't we just start with the cat's sensible looking diagram? Why all the rhombuseseses?"

"Well, if I just drew Nero's diagram, it would look like magic. This way, you get to see how the changed speed follows naturally from the changes in space and time that we've already talked about."

"Yeah, I guess that's right. Seeing how you get the speed from the rhombuseseses does make things clearer."

"You know, if you have that much trouble with 'rhombuses,' you could always go with 'rhombi' instead."

"Oh, it's not a problem. All those esses are fun to say: Rhombusesesesesesesesses!"

Figure 6.3.

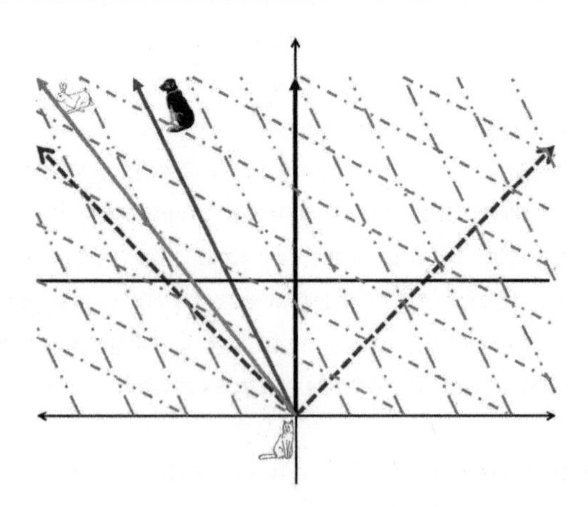

The discovery that Nero sees the bunny's speed as less than the speed of light is surprising—and an example of the general rule that no object can ever move faster than the speed of light. The speed of the bunny as measured by Nero is faster than that measured by Emmy, as we would expect, but slower than we would expect according to the Galilean rules. This will be true no matter what the speeds of the moving object and moving observer are: when we determine the speed of an object seen by a moving observer, that speed is always less than the speed of light. On our spacetime diagrams, the grid representing the position and time according to the moving cat will squeeze down toward the 45-degree worldlines for light rays, leading to higher and higher observed speeds for the object, but no amount of distortion will ever make the speed equal to the speed of light. No object that one observer sees as moving at a speed less than the speed of light will ever be seen as moving at or above the speed of light by any other observer, no matter how fast that second observer may be moving relative to the first observer. This result is one of the reasons why people say that relativity establishes the speed of light as an absolute upper limit on the speed of any moving object.

Just as any object that one observer sees as moving slower than the speed of light will be seen as moving slower than the speed of light by all other observers, anything one observer sees as moving at the speed of light will always be seen as moving at the speed of light by any other observer, no matter what the speed of that observer is relative to the first. This is exactly what we need to be consistent with the principle of relativity—since light moves at the speed of light, and the speed of light must be the same for all observers, anything that one observer sees moving at exactly the speed of light must be moving at exactly the speed of light for all other observers. Light is the only thing known to modern physics that moves at the speed of light, but should some exotic particle be discovered that moves at the speed of light, it will always be seen to move at the speed of light, no matter who observes it.

"What if it moves sideways?"

"What do you mean?"

"Well, when you talked about measuring stuff back in Chapter 1, distances at right angles to the moving observer didn't change. So, you just need to find something that's moving sideways at the same time as it's moving toward the observer. The sideways velocity plus the straight ahead velocity can be more than the speed of light."

"That's a good catch, and it brings up something else that's different about relativity. According to the Galilean rules we talked about back in Chapter 1, the perpendicular components of the velocity don't change, but in relativity, they do."

"But, wait, the distances don't change in the perpendicular direction. So how can the velocity change?"

"Velocity isn't just about distance, it's also about time. The perpendicular distances don't change—the length contraction we talked about earlier only happens along the direction of motion—but the *time* changes. Moving clocks run slow, so the moving observer sees the particle cover the same sideways distance in a different amount of time. A moving observer sees a lower sideways velocity, with the end result that the total speed, including all three dimensions, is always less than the speed of light."

"But, wait, doesn't that mean changing the direction?"

"That's right. Objects that are moving perpendicular to the observer's direction of motion bend toward the direction of motion."

"How much do they bend?"

"Well, if Nero is moving at four-fifths the speed of light, and he fires a rocket off at four-fifths the speed of light perpendicular to his motion, he sees the rocket going straight up, always at the same left-right position. You would appear to him to be moving from right to left, and for every unit you move left, the rocket moves one unit up."

"Okay, that makes sense."

"Now, under the Galilean rules, you would see the rocket moving off at a 45-degree angle—moving at four-fifths the speed of light in the horizontal direction and four-fifths the speed of light in the vertical direction."

"Yeah, that's pretty obvious."

"In relativity, though, the rocket's perpendicular speed is only 12/25ths the speed of light (0.48 instead of 0.8). The angle between the direction of the rocket and the direction of Nero's motion would only be 31 degrees. For each unit he moves to the right, you see the rocket move only half as far up" (see Figure 6.4).

"That's kind of strange . . ."

"True, but that's how relativity works. This also keeps the total speed under the speed of light—if the rocket moved off at 45 degrees in your frame, its total speed would be 1.13 times the speed of light. Under the relativistic rules, though, you see it moving at only 0.93 times the speed of light."

"And this is because his clock ticks at a different rate than mine?"

"Exactly. The two of you are moving relative to one another, which means your clocks tick at different rates. So even though you agree about distances in the perpendicular direction, you disagree about speeds because your clocks tick at different rates."

"That's your explanation for everything, isn't it? It's always because of timing."

"Well, yeah. Because it is all about time—N. David Mermin even titled his excellent book on special relativity *It's About Time*, because you can explain all of the effects of relativity through the effect on time. He has a

Figure 6.4.

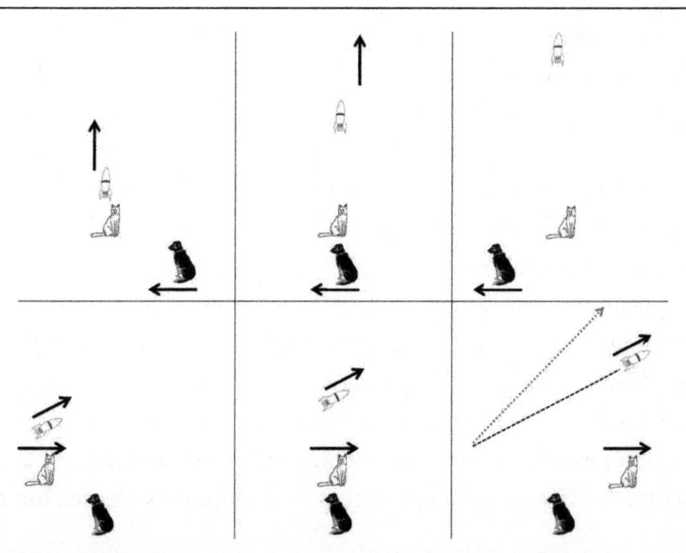

really cool example, too long to copy for this book, showing how you can reproduce all of special relativity—time dilation, length contraction, velocity addition, and all—using slow-moving trains of clocks that are out of synch with one another in the same way that clocks in different places are out of synch with one another according to relativity."

"Yeah, but why would you do that?"

"Well, it's a nice illustration of the importance of time and clock synchronization. You could use it to give somebody the sensation of what it would be like to move at speeds close to that of light. Or, you know, it could be a really complicated and geeky practical joke, to mess with somebody's head."

"Hmmmm . . . Could we build two slow-moving trains of cats in the backyard, so I could mess with their heads?"

"I don't think we have the room. You should stick to just reading about it in books."

"You're no fun at all, you know."

YOU CAN NEVER PUSH HARD ENOUGH: MOMENTUM AND FORCE

The process of converting velocities from one frame to another explains why an object that one observer sees as moving slower than light speed will be seen as moving slower than light by any other observer, but it does not rule out the possibility of something moving faster than light in the first place. An ambitious dog with access to rocket technology might very well think, as a result, that if she just pushes something hard enough, she can get it up to a speed that is faster than light for *all* observers. This approach is doomed to failure, though, because of the relativistic definition of momentum.

Noether's theorem tells us that because the laws of physics are symmetric in space—they work the same way in Schenectady as in Tokyo—there must be some invariant physical quantity that remains the same as an object moves from one place to another. We call this quantity the momentum, and if you calculate the total momentum of a collection of interacting objects, that total does not change as they move around, though the momenta of the individual objects may change due to their interactions.

As any dog who has watched sports on television knows, the usual definition of momentum is mass times velocity. A slow-moving, heavy dog, like a bloodhound, has the same momentum as a fast-moving light dog, like a hyperactive Jack Russell terrier.

Momentum is most important in the study of collisions, where knowing that the total momentum of the colliding objects—two tennis balls bouncing off each other, say—is the same after the collision as before the collision dramatically simplifies the process. All we need to do is redistribute the total among the two balls in a manner consistent with the rules of physics, which doesn't require detailed knowledge of what happens while they're in contact. This leads us, for example, to the conclusion that two objects of equal mass colliding head-on will exchange velocities, which we found in Chapter 1 by changing to a moving frame of reference in which the collision looked simpler, then going back to the original frame using the Galilean rules for converting velocity measurements. This collision clearly gives us the same momentum before and after: if you have one ball moving at 5 m/s and the other at 10 m/s before the collision, you still have one ball moving at 5 m/s and another at 10 m/s after they collide; which ball has which speed has changed, but the total momentum of the pair is the same.

The fundamental symmetry in space that Noether's theorem tells us leads to momentum conservation is the same in relativity as in classical physics, so momentum should still be an invariant quantity. The method used to calculate it must change, though, as we see from the second example discussed in Chapter 1, a heavy moving object colliding with a lighter one moving in the opposite direction. In that case, we found that the lighter object should bounce back with very nearly twice its initial speed. This can't work for objects moving faster than half the speed of light, though, since it would require the lighter object to move faster than light.

You might think that the solution is just to replace the Galilean rules for converting velocities between frames, but that doesn't work. Even if you use the proper relativistic rules for velocity conversion as you move between frames, you find that after the conversion, the total momentum, calculated from the usual mass-times-velocity formula, seems to change during the collision, which shouldn't happen. In order to maintain the idea of **conservation of momentum**, then, we need a new definition of momentum for objects moving at relativistic speeds.

"Let me guess: multiply by the Lorentz factor, γ."

"Not exactly, but something very similar. Good catch."

"It's really pretty obvious, dude. I mean, it turns up everywhere."

To construct a relativistic version of momentum, we need to look at the classical definition—mass multiplied by velocity—and replace the ordinary velocity with some more general velocity that both a dog at rest and a moving cat can easily determine. Velocity is the distance moved by an object divided by the time required to move that distance. The appropriate distance for a relativistic momentum is easy: it's the spacetime interval Δs, because all observers will get the same Δs. The appropriate time is less obvious, because different observers disagree about the timing of events, but there is one "special" frame whose time interval can easily be calculated by any observer: the frame in which the object is at rest. If we want to know the time that passes according to a cat moving with the object, we simply divide the time measured by a stationary dog by the Lorentz factor, γ, for the moving cat.

The resulting spacetime velocity can be expressed as a single number, but it's more useful to break it down into four parts: three spatial components (corresponding to velocity in the north-south, east-west, and up-down directions of space) and one time component (which we will talk more about in the next chapter). We use the three spatial components to define momentum in relativity.

The proper definition of momentum, good for any speed, is thus

$$p = \frac{1}{\sqrt{1 - v^2 / c^2}} mv$$

This looks like the classical momentum multiplied by the Lorentz factor, γ, and it is frequently written in that way, but there's one subtle difference. The speed that goes into the Lorentz factor we've used before is the speed of a moving *observer* looking at some event of interest. The speed that goes into the definition of momentum is the speed of the *object* according to the observer calculating the momentum, not the speed of the canine observer herself. If you have several different objects moving at high speed, each will have its own speed-dependent factor.

"Doesn't that make it harder to work with?"

"Relativistic momentum is much more complex mathematically, that's for sure. It's the correct definition, though, and will work at any speed."

"Yeah? Then why didn't anybody notice it before Einstein?"

"Just like with time and position measurements, the factor is very, very close to one for any speed you're ever likely to encounter. You need an object to be moving at about 14 percent of the speed of light before its momentum differs from the classical prediction by even 1 percent."

"In that case, who cares?"

"Particle and nuclear physicists care. For the Large Hadron Collider, which accelerates protons to 99.9999991 percent of the speed of light, the momentum of a proton is about seven thousand times what the classical formula would predict."

"That's . . . lots bigger. Lots and lots bigger."

"Exactly. Collisions of particles moving at that kind of speed look very different from collisions of slow-moving tennis balls. You need relativistic momentum to understand what's going on, and every particle accelerator ever made has behaved exactly as predicted by the relativistic momentum we just defined."

"OK. I still don't see how this limits me to moving slower than light, though."

"Well, keep reading, because that's what's next."

The momentum of a moving object is the truly essential quantity for determining how a moving object will behave in response to some interaction with another object. Whenever a dog interacts with some object—pushing a ball along the floor with her nose, pulling on a leash, or grabbing a toy in her teeth and shaking it vigorously—the principal effect of her actions is to change the momentum of that object.*

Most humans, and many dogs, have run across Isaac Newton's laws of motion, mentioned briefly in Chapter 1:

> *First law:* An object in motion tends to remain in motion in a straight line at constant speed unless acted on by an external force.
>
> *Second law:* The force needed to produce a particular acceleration is equal to the acceleration multiplied by the mass of the object (usually written as the equation $F = ma$).
>
> *Third law:* For every action there is an equal and opposite reaction.

These three rules, put forth by Newton in 1687, represent the birth of physics as a mathematical science and form the core of classical physics. The second law in particular is the key to understanding the motion of objects: if you know how the object interacts with the rest of the world, you can calculate the forces on that object, and from those forces, you can determine the acceleration.

While it may seem so based on the common formulation, Newton's second law is actually a statement about the momentum. "Acceleration" just means "change in velocity," so the force exerted is the mass times the change in velocity. Classically, momentum is mass times velocity, so force is really a measure of the change in an object's momentum** (provided the mass does not change).

---

* Or at least try to change the momentum, in the case of the leash.

** Newton himself originally presented the second law as a statement about momentum rather than acceleration. The original Latin is "Mutationem motus proportionalem esse vi motrici impressae, et fieri secundum lineam rectam qua vis illa imprimatur," translating to "The alteration of motion is ever proportional to the motive force impress'd; and is made in the direction of the right line in which that force is impress'd." Newton consistently uses "motion" to mean what we now call "momentum," so the primary expression of the second law is in terms of momentum. The $F = ma$ formulation is a simplification used in introductory physics classes to avoid having to use calculus.

While relativity changes the way we think about momentum, this central idea carries over: the force needed to change the speed of a particle is related to the change in the momentum of that particle. We find the force needed to produce a given acceleration by applying this principle to our definition of relativistic momentum.

"And the force needed to accelerate something is the usual force multiplied by the Lorentz factor, γ. Let's move this along, shall we?"

"That's the obvious guess, because γ turns up all over the place in relativity, but it's more complicated than that. The momentum involves *two* things that depend on the speed: the velocity of the object and the Lorentz factor. The exact relationship between force and acceleration for a relativistic particle depends on both, as well as on the direction of the force relative to the direction of motion. For a force that is strictly in the direction of motion, such as a rocket trying to accelerate a ship up to the speed of light, the force is actually the mass times the acceleration times the *cube* of the Lorentz factor, γ."

"Why is that?"

"You need calculus to explain it in detail."

"Forget I asked, then. I don't like calculus."

"It's really not that bad, but it's also not important. The important thing is that the force required to accelerate an object increases dramatically as the speed of that object increases."

"Which was what I was going to say. So, this was just some pedantic physics professor correction, wasn't it?"

"Yeah, I guess it was."

When we determine the force needed to produce a given acceleration using our definition of relativistic momentum, we find that the force needed to produce a given acceleration increases dramatically as we get closer to the speed of light.

A running dog reaches speeds a bit higher than 20 m/s, and if it takes the dog two seconds to reach full speed, that's an acceleration of 10 m/s per second. A 20 kg dog can obtain this acceleration by exerting a force of 200N, which is about equal to her weight due to the Earth's gravity. If that dog wanted to get the same acceleration starting at half the speed of light, she would need to exert 307N of force, half again her weight. At four-fifths the speed of light, the force increases to 925N, four and a half times her weight. At 90 percent of the speed of light, it's 2,400N, and so on.

The closer you get to the speed of light, the more force is needed to keep accelerating. This makes it impossible to ever accelerate an object from speeds below the speed of light up to the speed of light. As the speed gets higher and higher, the force required to gain that next little bit of speed becomes larger and larger. Actually reaching the speed of light when accelerating from rest would require an infinite amount of force, which no real engine can ever provide.

"That's just if you want to get there quickly, though, right? I mean, if you don't mind accelerating more slowly, then you can do it with a finite force, right?"

"In a really abstract sense, I suppose, but you're talking a really long time."

"I'm a patient dog!"

"That's news to me. I mean, you get all antsy and start jumping around if it takes me more than a second to open the treat jar."

"There's no need to get personal, dude. Just answer the question: can I use a finite force to reach the speed of light if I'm patient or not?"

"For an extreme value of 'patient,' maybe. A 200N force, at 90 percent of the speed of light, gets you an acceleration of 0.83 m/s per second. If you could maintain that acceleration, it would take about 36 million seconds to get that last 10 percent, or a bit more than a year. And, of course, you wouldn't really get to maintain that acceleration—at 99 percent of the speed of light, the acceleration from 200N would be 0.0016 m/s, and it would take 1.9 billion seconds, about sixty years, to cover that last 1 percent. And the acceleration keeps getting smaller the faster you go."

"So . . . that's a no, then?"

"That's a no. Any finite force takes an infinite amount of time to accelerate an object up to the speed of light. And nobody is that patient."

"Especially not a dog. So much for that idea, then."

ARRIVING BEFORE YOU LEFT: CAUSALITY AND FASTER-THAN-LIGHT TRAVEL

We have established that nothing that any one observer sees as moving slower than the speed of light will ever be seen as moving at the speed of light by any other observer, and it is impossible to accelerate any object up to the speed of light. These two items pretty conclusively rule out travel at or above the speed of light, but a particularly lawyerly dog* might argue that this leaves a loophole: we have only shown that it's impossible to use ordinary forces to reach speeds greater than that of light. Some magic technology, using physics beyond what we understand, might start a stationary object moving faster than light in some manner that does not involve accelerating through all the intervening speeds.

Faster-than-light drives using unexplained magic technology are a staple of science fiction stories involving spaceships, but this, too, is forbidden in our modern understanding of physics. If faster-than-light travel were possible, this would produce serious problems with causality, the principle stating that causes always precede effects.

We can see the problem with faster-than-light travel by considering a couple of spacetime diagrams. Let's imagine that Emmy the dog and Nero the cat are going about their business, as in previous diagrams, when a space alien flies past in a UFO moving at four times the speed of light.

In Figure 6.5, the diagram on the left shows this set of events with a grid representing Emmy's position and time measurements. The UFO moves from left to right across the diagram, passing first Emmy (at the spacetime position marked 1), and then Nero (at the spacetime point marked 2) about half a unit of time later. There's nothing odd or disturbing about this diagram by itself.

The diagram on the right, though, shows the same events overlaid with the slanted grid representing Nero's measurements of position and time. The dot-dash lines represent instants of time in his frame, and we see that

* A legal beagle, as it were.

Figure 6.5.

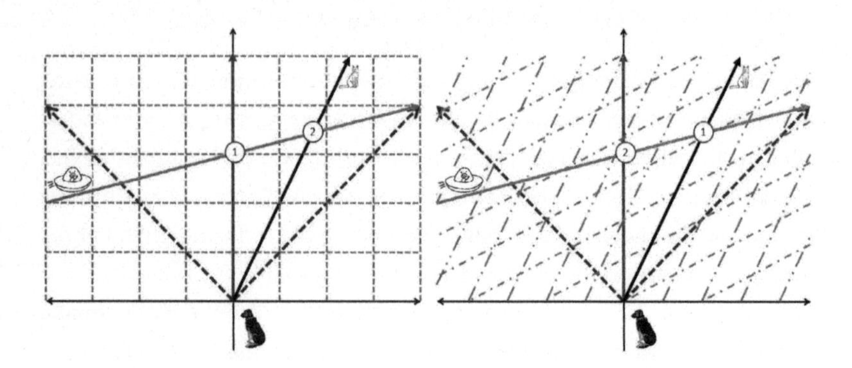

according to him, the order of events is reversed: the UFO passes him first and then Emmy about half a unit of time later. The UFO that Emmy saw streak past from left to right moves from right to left in Nero's frame.

This might not seem like such a big deal. Even under the Galilean rules from Chapter 1, moving observers can disagree about the direction of motion of an object, and in Chapter 5 we showed that moving observers will disagree about the ordering of events with space-like separation. This situation brings up a more fundamental problem, though, because the alien is present at both of these events, providing a causal connection between them.

So, let's imagine that rather than simply passing through, the alien helps Emmy play a funny trick. At point 1 in Emmy's diagram, the alien accepts a water balloon from her, and at point 2, it drops the balloon on Nero. An observer at rest relative to Emmy sees the alien take the balloon, then soak the cat, which is hilarious.

An observer moving with Nero, on the other hand, sees him get soaked *first* and only later sees the alien get the balloon from Emmy. While this might still be funny, it would also be thoroughly baffling. The cause of Nero's soaking happens only after he has been drenched. This reversal of causality would make it nearly impossible to construct a coherent picture of the universe and is the most fundamental reason why the speed of light is the ultimate limit for moving objects.

"So, let's see . . . According to the cat, the alien is moving backwards, so time is reversed. So, he would be wet, then the alien would come along and collect the water into a balloon, then hand it to me?"

"No, it'd be weirder than that. The alien would fly in backwards, soak Nero, and also carry a full water balloon backwards to you, but you would have your own balloon. When you met, both your balloon and the one carried by the alien would vanish. Nero would still be dry before the alien arrived, and wet afterwards, while you would have a balloon up until the alien arrived, and then not after that."

"But, wait, if the cat was dry before the alien arrived and wet afterwards, where did the water come from?"

"That's exactly the problem. The water just sort of appears out of nowhere to soak Nero but also forms into a balloon that the alien is carrying."

"But that doesn't make any sense!"

"Right. Which is why you can't have faster-than-light travel: it makes a mess of causality, and if cause doesn't precede effect, you can't construct a coherent history of the universe."

"So, my plan for a superluminal water-balloon cannon is right out?"

"For many reasons, but the most fundamental reason is the problem of causality."

"Now, this rules out sending stuff faster than the speed of light, but how about messages? Information isn't really stuff, so maybe I can just send messages faster than the speed of light."

"That doesn't work, either, for the same reasons of causality."

The space alien thought experiment clearly rules out sending physical objects like water balloons from one place to another, but our hypothetical canine lawyer might argue that this doesn't rule out sending nonphysical messages faster than light, even if it's impossible to send physical objects at those speeds. But the sending of messages faster than light is also forbidden by relativity for much the same reason that sending physical objects is: if superluminal communication is possible, you can send messages into the past, which can then violate causality. We can illustrate this with another spacetime diagram, this time using two pairs of dogs: Harley the poodle and Truman the Boston terrier sitting at rest, while Anson the Lab mix and Bodie the yellow Lab run past at high speed, each carrying a device allowing instantaneous communication with the other dog in its pair.* Figure 6.6 shows the diagram for the situation.

Here's the scenario for violating causality with instantaneous communication: Harley, at the central position in the diagram, mistakes Anson for a cat and splats him with a water balloon at the point marked 1 on the diagram. Anson then uses his instantaneous communicator to send a message to Bodie saying, "Hey, I just got soaked. Tell him not to do that!" As a particular instant according to the moving dogs falls on a slanting line, this message travels down and to the left in the diagram.

Bodie passes the message on to Truman at the point marked 2 on the diagram. He then uses his instantaneous communicator to send a message to Harley saying, "Hey, don't throw your water balloon!" That message arrives at the point marked 3, which is *before* event 1 according to Harley and thus in time to prevent the initial soaking. But if Anson didn't get wet and send a message, then there never would've been a message sent telling Harley not to throw the balloon, in which case Anson would've been soaked. In which case the message would've been sent and the soaking prevented . . . And so on, round and round in an endless loop.

So, even sending messages faster than the speed of light creates problems with causality. Faster-than-light communication effectively sends a message into the past, which is enough to create a paradox. For this reason, nothing—not an object and not a signal—can be sent faster than the speed of light.

"Yeah, but that's just because you have the instantaneous communication working in the wrong way."

"What do you mean?"

---

* Instantaneous communication is the most extreme version of faster-than-light communication. The same argument works for communication devices that sent signals faster than light but not instantaneously; it's just easier to draw the diagrams for the instantaneous version.

Figure 6.6.

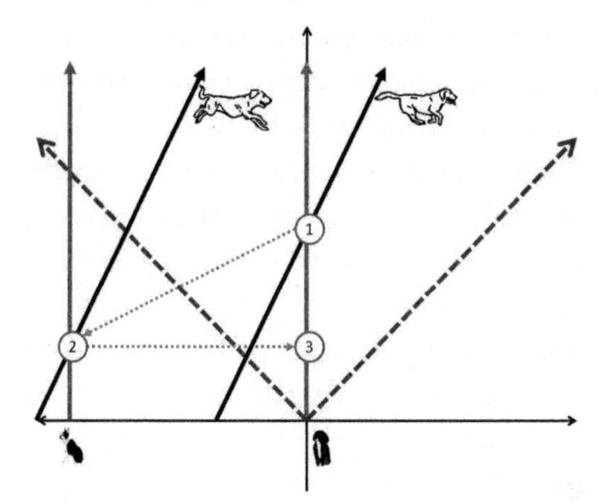

"Well, it sends messages into the past because you have the moving dogs sending messages on a diagonal line. If they sent their message on a horizontal line, there wouldn't be a problem."

"Well, there wouldn't be a problem with causality, but there would still be a problem with relativity."

"Why?"

"Well, if the 'instantaneous' messages only move on horizontal lines in this diagram, that would mean that they're only instantaneous to the stationary dogs. Moving dogs would see their messages take some time to travel from one place to another. But that would mean that the instantaneous communicators worked differently for the moving dogs than for the stationary dogs. And that violates the principle of relativity."

"Oh. OK. But wait, you're saying that nothing can move faster than light, right?"

"Right."

"OK, but how about this: when you shine a laser pointer on the floor and I chase it, the spot on the other side of the room moves faster than your hand does, right?"

"Right."

"And the farther the spot is away from you, the faster it moves, right?"

"Right."

"So, if I shoot a laser really far, like, to the moon, and move it from side to side, the spot will move faster than the speed of light. So, stick that in your pipe and smoke it, Einstein! Stuff can too move faster than light!"

"You're right that the laser spot would move faster than the speed of light as it crossed the surface of the moon, but it's not a problem for relativity."

"Yes it is. You said nothing could move faster than the speed of light."

"No, I didn't. I said no *signal* can move faster than the speed of light. The laser spot moves faster than the speed of light, providing you sweep it across the full width of the moon in about a hundredth of a second, but a laser spot isn't a physical object, and it can't be used to send a signal."

"Sure it can. You just wiggle it back and forth, and send your signal in Morse code."

"You can send a signal in Morse code, but that signal isn't encoded in the position of the spot on the moon; it's encoded in the blinking of the light seen by somebody on the moon's surface. That pattern of blinking is carried along the light itself, which moves at the speed of light, by definition."

"But the spot goes from one side of the moon to the other faster than the speed of light."

"Yes, it does. And people with really good clocks and light detectors would confirm that the laser spot tracked across the moon faster than light, later on, when they got together to compare measurements. But the spot moving across the moon doesn't convey any information from one side of the moon to another, so it's not a problem. You can't use it to send a message from one side of the moon to the other without going back to Earth first, at the speed of light. Causality is perfectly safe."

"OK, I guess."

"You're far from the only one to be tripped up by this, though. The distinction between sending information and just seeing something appear to move faster than the speed of light is a subtle one and hard to keep straight."

"That's a relief, I guess. But I still want to be able to go faster than light. Think of all the bunnies I could catch!"

"Sorry, but you can't always get what you want. But, you know, if you try sometimes, you just might find you get what you need."

"Hey, that's clever. You should write that down."

We have shown that the speed of light is a fundamental limit on the motion of objects or information in three different ways: by showing that any slower-than-light velocity will always be seen as slower than light, no matter how fast the observer is moving; by showing that accelerating a massive object to the speed of light would require an infinite force; and by showing that faster-than-light travel or communication leads to violations of causality. These are practical arguments as to why the speed of light is unattainable, but light speed as a cosmic limit is ultimately a fundamental property of our universe. The relationship between space and time that we explored in Chapter 5 demands that there be some maximum speed for moving objects and that this speed be the same for all observers. From Maxwell's equations, we know that the speed in question is the speed of light.

These results are a mixed blessing for humans and dogs encountering relativity for the first time. On the one hand, limiting the sending of objects and information to speeds less than the speed of light means that causality is preserved, providing another island of stability amidst the changes of relativity: as weird as relativity gets, you never see effects happening before their causes, so at least some common sense is preserved. On the other hand, though, it means that travel to distant planets will necessarily require vast amounts of time because, in the words of Douglas Adams, "Space is big. You just won't believe how vastly, hugely, mind-bogglingly big it is."* To reach the nearest star would take four years traveling at the speed of light; at the speed of Voyager I, the fastest-moving man-made object in history, it would take more than seventy thousand years.

Generations of science fiction writers, in books and on television, have invented numerous imaginary ways of circumventing this light-speed

---

* See Chapter 8 of Douglas Adams's *The Hitchhiker's Guide to the Galaxy*, originally published in 1979.

limit. Relativity is remarkably unforgiving, though, and as far as we know, there is no way to move through space faster than the speed of light. We just have to accept light speed as an upper limit and move on to discuss the world's most famous equation.

"$E = mc^2$!"

"Yes, that's right."

"Oh, good. That's the coolest part of relativity—turning mass into energy. I hope it involves eating stuff."

"Again, that has more to do with biology than physics."

"Yes, but isn't everything ultimately just applied physics?"

"Sure, but you don't need relativity to understand how eating food gives you the energy to run around chasing things and barking. Relativistic energy is really only relevant for extremely fast-moving objects—generally, subatomic particles."

"And those aren't much fun to eat."

"No, they aren't."

"Hey, you know, you left one big loose end hanging in this chapter."

"Did I?"

"Yeah. When you defined momentum, you talked about using the spacetime interval to find a general velocity to use for momentum. But that would have four components, and you only talked about using the three spatial parts."

"That's right."

"So, what happens with the fourth part, the time bit?"

"An excellent question, that. And very timely, pardon the pun."

"Oh. Let me guess, it's important for $E = mc^2$."

"You've got it."

"And that's the next chapter, so I should stop talking and turn the page, already."

Chapter 7

A DOG IS ALWAYS IN MOTION: $E = MC^2$

I NOTICE THAT THE BACK DOOR IS AJAR, so I step past Emmy's bowl to push it all the way closed. Hearing me in the mudroom, she comes racing over to check her bowl for food.

"There's nothing new in there," I say. "You've already had your breakfast, and dinner isn't for a few hours yet."

"Awww . . . ," she says, looking pathetic. "But I need more food. I'm *starving.*"

"You are not starving. In fact, at your last checkup, the vet said you needed to lose a few pounds."

"Oh, what does some silly human vet know? He doesn't know what it's like to be a starving dog."

"I think I trust his education more than your appetite. You get all the kibble you need."

"I do not. It doesn't give me anywhere near the energy it should, according to physics."

"How do you figure?"

"Well, relativity tells us that $E = mc^2$, right? So, if I eat a kilogram of kibble, I ought to get 90,000,000,000,000,000J of energy. That's enough for me to run at 10 percent of the speed of light, but I can't even break the speed of sound." She sulks. "Either it's defective kibble, or physics is wrong."

"Both physics and your kibble are just fine. First of all, I don't think you've ever eaten a kilogram of kibble at one sitting, except maybe that one time when the bag broke open on the way in from the car." She wags her tail at the happy memory. "More importantly, though, you can't possibly expect to get that much energy out of the food you eat. The chemical bonds responsible for the energy content of your food account for a miniscule fraction of the total mass. One hundred grams (g) of your kibble contains about 420 food calories, which is 1.76 million joules of available energy. The mass energy of that kibble is 8.99 quadrillion joules, so you're getting 0.00000002 percent of the mass as energy."

"See, I told you, my kibble is defective."

"It's not your kibble. Even high-calorie-density foods aren't much better. For instance, 100g of bacon provides about 588 food calories for the same mass energy, so you're only getting 0.00000003 percent of that—if we gave you that much bacon, which we're not going to do, so stop drooling."

"So, what you're saying is, I need a better way of extracting energy from my food."

"Even if you were a nuclear-powered dog, you wouldn't get more than a fraction of a percent of the total mass out as usable energy, so forget it."

"I don't know about that, dude. Those numbers sound kind of goofy to me. If there's all this energy in my kibble, how come I can't get it out? I think it's your physics that's wrong."

"I highly doubt that. Relativity is extensively tested, and there's no doubt that mass and energy are equivalent. Even for bacon and kibble."

"But it doesn't make any sense! Why would a bunch of kibble have all that energy? It's just sitting there! And why would the speed of light come into it at all?"

"It's not just sitting there, it's moving through spacetime at the speed of light."

She cocks her head sideways and looks at me like I sprouted an extra head. "Dude, I think *you* may need to go to the vet. I can see my kibble just fine, and I assure you that it's not moving."

"It's not moving through *space*," I say, "but remember, in relativity, you can't talk about space and time as independent things. And it's definitely moving through time."

"It is?" She looks puzzled.

"Sure. For every second that passes, it moves one second into the future."

"Oh, *that*. I thought you were saying that my kibble had a time machine or something. Of course it's moving into the future at one second per second. Everything is. So what?"

"That's exactly the point. Everything is moving into the future at one second per second, which means that everything is moving through spacetime. As a result, everything has energy associated with that motion, with the total energy given by the mass multiplied by the speed of light squared: $E = mc^2$."

"So, if everything's moving all the time, why don't I notice this?"

"You do, it's just that at everyday speeds, the effects of relativity are small, and we perceive space and time as different things. You experience motion through space as visible motion, while motion through time you perceive as just the passage of time. You need to combine the two to understand energy in relativity, but once you do, the origin of the **rest energy** is clear."

"Hmmm . . ." She thinks for a minute. "That's an interesting story. I'm not sure I believe it, but it's definitely something to think about."

"Believe it. It's well-established modern physics."

"OK. But, you know, between the racing through spacetime and the thinking of deep thoughts . . . well, those things really work up an appetite, you know?"

I sigh. "Dinner's not for hours yet, but if you're willing to practice a few tricks, I'll give you a couple of treats."

"Oooh! Treats are *way* better than kibble!" She does the happy dance all through the kitchen while I walk over to the treat jar.

Einstein's name is forever associated with the equation $E = mc^2$, arguably the best-known scientific equation in history.* The equation is also forever associated with both nuclear power and nuclear weapons, conjuring images of either clean energy or mushroom clouds, depending on whom you ask.

This is a little unfortunate, both for Einstein and for the equation itself. While it is undeniably important for physics, $E = mc^2$ is only a small piece of relativity, and not even the most important piece. And nuclear energy and nuclear weapons are only a small piece of the meaning of $E = mc^2$.

In this chapter, we talk about the real meaning of the world's most famous equation and show how it is a natural consequence of the spacetime concept introduced in Chapter 5. First, though, we need to talk about energy, the E in $E = mc^2$, and how it came to be one of the most important concepts in physics.

FROM LIVING FORCE TO TIME SYMMETRY: THE HISTORY OF ENERGY

Like many great ideas in physics, the notion of energy was first encountered as a mathematical accident. Gottfried Wilhelm Leibniz, the great German scientist and scientific rival of Isaac Newton,** noted that a particular combination of properties, the mass multiplied by the square of

---

* David Bodanis even called it that in the subtitle of his book *$E = mc^2$: A Biography of the World's Most Famous Equation* (New York: Berkley Trade, 2001).

** Newton and Leibniz each independently invented calculus at more or less the same time—Newton was probably first, but being a secretive sort (he also dabbled in alchemy), he didn't publicize his discovery until after Leibniz had begun to circulate his results. This kicked off a lengthy and acrimonious dispute over which discovery took priority, which in turn became further entangled with nationalism—British scientists used only Newton's results and disparaged those of Leibniz, while scientists on the European continent favored Leibniz. It has been claimed that this explains the relative dominance of German and French mathematicians in this era—the system of notation used by Newton was more cumbersome than Leibniz's and held British mathematicians back.

the speed, turned up again and again in calculations about the motion of objects. He dubbed this combination the *vis viva* (Latin for "living force") and proposed that it was a fundamental quantity, like momentum. Leibniz's method only worked for a few simple problems, though, so it was not widely adopted.

With the invention of steam engines and the beginning of the Industrial Revolution, physicists and engineers began to study the behavior of complicated systems in more detail and eventually realized that a slight modification of Leibniz's *vis viva* is one of many forms of what we now call energy.* The total energy of a system of objects does not change in time, but the energy can be shifted between many different forms. An object in motion has **kinetic energy.** An object that has the potential to start moving (say, because it is on a high shelf, ready to fall if a dog should bump into it) has potential energy due to its interaction with other objects. A hot object contains thermal energy due to the random motion of the atoms making it up, and so on.

Many physical processes can be understood using the principle of **conservation of energy.** The bouncing of a tennis ball thrown for a dog to chase, for example, can be understood in terms of energy flowing between different forms. As it bounces up from the ground, the ball has kinetic energy. At the highest point of a bounce, that kinetic energy turns into potential energy due to the gravitational interaction between the ball and the Earth. As the bounces become shorter and eventually stop, the initial kinetic and potential energy turns into thermal energy, slightly increasing the temperature of the ball and the ground.

The most important form of energy for our purposes is the kinetic energy associated with an object in motion. This is the direct descendent of Leibniz's *vis viva*, and at low speed it is given by the formula: $\frac{1}{2}mv^2$

The kinetic energy of an object is one-half its mass times the speed squared. The kinetic energy increases more rapidly than the momentum, so a hyperactive miniature schnauzer can have substantially more energy than a slow-moving basset hound, even though the hound may outweigh the schnauzer by quite a bit.

* The English name "energy" was introduced by the British physicist Thomas Young in the early 1800s.

Kinetic energy is useful for understanding a number of everyday phenomena. For instance, imagine two dogs of the same breed and equal mass, one running at twice the speed of the other. When they decide to stop, they must turn their kinetic energy into some other form of energy—generally, thermal energy in their leg muscles and the ground they're running on—meaning that the faster-moving dog will require four times as much distance to stop, because she has four times the kinetic energy of her littermate. This applies to all fast-moving objects, including human constructs like cars, which is yet another reason why it's a good idea for dogs to stay out of the path of oncoming cars—a heavy, fast-moving object like a car has a lot more kinetic energy than any animals a dog would have experience with, and it will have great difficulty stopping short of a dog in the road.

"What about an elephant?"

"What?"

"You said that a car has more kinetic energy than any animal, but an elephant is way bigger than a car."

"I said a car has more kinetic energy than any animal a dog would have experience with. While a running elephant can have more kinetic energy than a car, I don't think you have any direct experience with elephants, do you?"

"I follow a couple on Twitter. Does that count?"

"No."

"Oh. Well, they do have dogs in Africa, you know. *They* have to worry about elephants. Some of them, anyway."

"OK, fine. A car has more kinetic energy than any animal a dog is likely to have experience with, except for dogs in places with elephants."

"And hippos. They're pretty big, too."

"Now you're just being silly. Anyway, even if you're talking about elephants, a car can have more kinetic energy, because the kinetic energy depends on the square of the speed. A 1,000 kg car moving at 30 m/s has the same kinetic energy as a 7,500 kg elephant moving at its top speed of around 11 m/s. And 30 m/s is about 67 mph, so highway speed."

"Or, you know, surface-street speed for you, when you got that ticket back in Chapter 1."

"Don't remind me."

The idea of the conservation of energy, and most of the forms of energy we now think about, was established by the mid-1800s through the work of scientists and engineers like Sadi Carnot, James Joule, William Thomson (Lord Kelvin), and Hermann von Helmholtz. The deeper physical significance of energy conservation was not fully appreciated until Emmy Noether proved her theorem in 1915, which showed that energy conservation can be seen as a consequence of time symmetry. The fact that the laws of physics are the same in the future as in the past implies that some physical quantity must also have the same value in the future as in the past; that quantity is the energy. Noether's theorem makes physical sense of the otherwise fairly arbitrary concept of energy and helps point the way toward understanding Einstein's most famous result.

$E = MA^2 \ldots E = MB^2 \ldots E = ??$: HISTORY OF THE WORLD'S MOST FAMOUS EQUATION

Einstein first put forth the equation $E = mc^2$ in the last of the four revolutionary papers he published in 1905.* In that paper, he arrives at the result through manipulation of the equations of relativity. He considers what happens when an object at rest emits light that carries a known quantity of energy and finds that the object's apparent mass decreases as a result.** This shows that mass and energy are different aspects of the same thing.

---

* The other three were a paper explaining "Brownian motion," a jittering motion of microscopic objects, which was a critical piece in sealing the acceptance of the atomic theory of matter; a paper explaining the photoelectric effect, which helped kick off the field of quantum mechanics; and the first paper on the special theory of relativity. Not bad for an unknown patent clerk.

** In the original papers, Einstein considers a "relativistic mass" that changes depending on velocity. That formulation is common in older books, but the modern convention is to think of the mass as a constant, measured by an observer at rest relative to the object, with the additional velocity dependence added to the definition of energy and momentum.

While Einstein's paper is mathematically impeccable, it fails to explain why mass is equivalent to energy. To get a feel for the fundamental meaning of the equation, we need the idea of a unified spacetime that we discussed in Chapter 5, our definition of relativistic momentum, and Noether's theorem.

In Chapter 6, we defined the spacetime momentum by dividing the spacetime interval Δs covered by some moving object by the time required to move through that interval as measured by an observer who is stationary relative to the object. This spacetime momentum has four components, corresponding to the three dimensions of space, plus one dimension of time. We call the three spatial components the momentum, which depends on the velocity of the object moving through space, but that leaves the fourth component, corresponding to motion through the time dimension.

At first glance, motion through time seems like an odd thing to be tracking, as, in the everyday world, everybody and everything moves into the future at the same rate of one second per second. That's an inherently classical view, though, because, as we have seen, observers moving relative to one another will disagree about the passage of time. What a stationary dog perceives as ten seconds of time may look like only seven seconds to a fast-moving cat. The two animals are thus moving through time at different speeds, and the rate at which they move depends on their velocity through space.

The fourth component of the spacetime interval is given by the time according to some observer, Δt, multiplied by the speed of light, c. To determine the time velocity, we divide this distance by the time according to an observer who is at rest relative to the object whose time velocity we are calculating, which is the time Δt divided by the Lorentz factor, γ. The two times are the same, so they cancel out, leaving the velocity through time as c, the speed of light multiplied by the Lorentz factor.

"Wait, wait, wait—you just spent a whole chapter telling me that nothing can move at the speed of light. But this speed you just calculated is bigger than the speed of light. Isn't that impossible?"

"Nothing can move *through space* at the speed of light. The speed *through time* can be greater than the speed of light, though."

"OK, but then, when I start moving through space, doesn't that mean I'm moving faster than the speed of light?"

"No. In fact, you move through *spacetime* at exactly the speed of light, no matter what your speed through space or time is."

"But how can you add two speeds together and get the same thing as when you only had one speed?"

"It's because spacetime distances add in a hyperbolic manner, like we saw in Chapter 5. When you add the speed through space and the speed through time, you square them both and subtract the time part. So, as your speed through space increases, your speed through time also increases, and the difference between those two components gives you exactly the same speed, namely, the speed of light."

"Wait, my speed through time increases? I thought moving clocks ran slow?"

"Moving clocks running slow is exactly what you get from increased speed in the time direction. If you get into a rocket ship and fly away at high speed, then back to where you started, you cover the same distance in spacetime as a different dog who just took a nap at the place where you started, but your clock says it took less time. So, when your speed through space increases, your speed through time also increases."

"OK, this is kind of freaky."

"It's different from what you're used to thinking about with ordinary velocity through space, that's for sure. But that's part of the deep structure of spacetime. The different rules for adding distances seem kind of strange compared to the ordinary Pythagorean theorem, but the end result has a certain mathematical elegance that's very pleasing."

"Pleasing to a physicist, anyway."

Multiplying the velocity through time by the mass to get the momentum, we have: $E = \gamma mc$

The time component of the momentum is just the mass multiplied by the speed of light and the Lorentz factor, γ. This quantity may seem a bit mysterious at first glance. It looks a little like the momentum but uses the speed of light, c, rather than the speed of the object, v. It still depends on the speed of the object, but only indirectly, via the Lorentz factor.

The interpretation of this mathematical object becomes clear when we look at it in terms of Noether's theorem. According to Noether's theorem, conservation of energy comes about because the laws of physics are constant in time, so it makes sense to identify the time component of the spacetime momentum with the energy. We need to multiply by the speed of light to get the units right (energy is mass times speed *squared*), and when we do, we get

$$E = \gamma mc^2 = \frac{mc^2}{\sqrt{1 - v^2 / c^2}}$$

The total energy of an object is equal to the mass multiplied by the speed of light squared and the Lorentz factor, γ.

The remarkable thing is that, unlike in the classical formula, energy does not go to zero when the object is at rest. When the object isn't moving, $v = 0$, and $\gamma = 1$. This means that the energy of an object that isn't moving is $E = mc^2$, the world's most famous equation. The speed of light squared is a huge number—about 90,000,000,000,000,000 m^2/s^2—so this means that everyday objects have a gigantic amount of energy simply by virtue of the fact that they have mass. This rest energy exists because even an object at rest in space—say, a dog dozing in a sunny spot—is necessarily moving through time at the speed of light, c. The energy described by Einstein's famous equation is the result of this motion.

"So, since time is just another dimension like space, motion through time produces energy just like motion through space does?"

"Exactly."

"How do you make all this weird stuff work with what we already know?"

"What do you mean?"

"Well, you said that the classical definition of kinetic energy as $\frac{1}{2}mv^2$ worked very well. That doesn't look anything like this relativistic thing, though. It doesn't have the speed of light in it anywhere or the Lorentz factor. So how can both of these things work?"

"Ah, that. The classical definition is just an approximation of the relativistic definition. Watch and see how this works."

CLOSE ENOUGH FOR CLASSICAL PHYSICS: KINETIC ENERGY AT LOW SPEEDS

Even for physics students, the first encounter with relativistic energy is a bit confusing, since the relativistic formula doesn't look like the classical one for kinetic energy at all. Somehow, these two very different formulae must be reconciled, given that the kinetic energy formula from classical physics works very well in most everyday circumstances. How does that work?

The answer is that the Newtonian expression for kinetic energy is a good approximation at low speed. The kinetic energy in relativity is the difference between the total energy γmc^2 and the rest energy mc^2, or $(\gamma - 1)mc^2$. This is very close to the classical value of $\frac{1}{2}mv^2$, provided the speed is small compared to the speed of light. How good is this approximation? Table 7.1 gives the kinetic energy calculated from the full relativistic expression and from the classical value for a 1 kg object moving at 300 m/s (the speed of a jet aircraft), 30,000 m/s (the speed of Earth in its orbit around the sun), 1 percent of the speed of light, and 10 percent of the speed of light.*

Table 7.1.

Speed	$(\gamma - 1)mc^2$	$\frac{1}{2}mv^2$
0.000001c (300 m/s)	45,000J	45,000J
0.0001c (30,000 m/s)	450,000,003J	450,000,000J
0.01c (3,000,000 m/s)	4,500,337,995,660J	4,500,000,000,000J
0.1c (30,000,000 m/s)	453,408,126,873,803J	450,000,000,000,000J

* The units of energy are joules, named after the British physicist James Joule, who did a famous experiment showing that thermal energy and kinetic energy were different forms of the same quantity. A 1 kg mass moving at 1 m/s has kinetic energy of 0.5J.

As you can see, the classical approximation agrees very well with the full relativistic expression. Even at 10 percent of the speed of light, it only differs from the full expression by a bit less than 1 percent; at the speeds a dog is likely to encounter, the difference is way too small to measure. For ordinary objects, the classical physics we know and love is enough. It's only when we start dealing with objects moving at speeds approaching the speed of light that we need to worry about the relativistic energy.

"OK, fine, you've shown that the kinetic energy formula from classical physics matches the formula from relativity, but you haven't addressed the really important thing."

"What's that?"

"The rest energy! I mean, relativity says that a 1 kg object has 90 quadrillion joules of energy even when it's not moving. That's not there in classical physics. So, if there's 90 quadrillion joules of extra energy around for every kilogram of mass, why don't we notice it? Who cares about 450 trillion joules of kinetic energy when you've got two hundred times that amount running around unaccounted for?"

"Well, the thing is, that energy is there, but it doesn't really matter for most ordinary experiments. When we deal with energy in the motion of ordinary objects, all we really care about is changes in the energy of the object. The mass energy is there, but it doesn't change forms during ordinary experiments. As long as it starts out as mass energy and stays mass energy, it doesn't change the motion of the objects, so we can just ignore it."

"I have a hard time with the idea that you can just ignore 90 quadrillion joules of energy."

"Maybe an analogy will help. Let's think about it in terms of money. The gross national product of the United States is about $14 trillion, give or take a bit. Compared to that, $20 is pretty insignificant, right?"

"Right."

"So, it doesn't matter if I spend another $20 buying you dog treats, right?"

"What?!?! No! I mean, yes! I mean, whichever involves you buying me treats!"

"But whether I spend the money on your treats or not, there's still this other $14 trillion running around unaccounted for."

"Yeah, but what *really* matters is that I get treats! I—oh, wait. I see what you did there."

"Right. The mass energy in everyday objects is huge, just like the US economy. But what matters for everyday physics problems is small changes in the energy, because those determine the motion of objects."

"Just like spending tiny quantities of money on my treats is what determines my happiness."

"Exactly."

"OK, I guess I can roll with mass energy. As long as I get my treats. Can we get to the fun part, now?"

"Which is?"

"Turning stuff into energy, of course! How do you convert mass into energy that I can use to watch TV and surf the Internet and zap bunnies with lasers?"

"Well, you—wait, what? Zap bunnies with lasers?"

"Did I say that out loud? Forget I said anything. Just talk about turning mass into energy. That *is* the next section, right?"

"Yes, it is. And when we're done with this section, we're going to have a little talk about the appropriate use of scientific equipment . . ."

THE POWER OF THE (STRONG) FORCE: FISSION, FUSION, AND NUCLEAR ENERGY

The strongest association most humans and dogs have with $E = mc^2$ is nuclear energy, for good or ill. Many people tout nuclear energy as a clean-energy technology, generating electricity without carbon dioxide, while others worry about the radioactive waste products generated by nuclear power plants. And even dogs worry about the danger of nuclear weapons in the hands of crazy humans.

There are two main processes humans have in mind when they talk about nuclear energy. One is nuclear fusion, which powers the sun and other stars: two light atoms, generally hydrogen, are stuck together to form a single, heavier atom, releasing some energy in the process. Nuclear fusion is the ideal clean-energy technology: it would generate energy from hydrogen gas, with the only waste product being helium gas, which could be used to fill balloons and make humans' voices sound funny. Unfortunately, despite decades of effort, no one has yet found a way to make commercially viable nuclear fusion.*

The other main nuclear process is nuclear fission, in which heavy elements like uranium or plutonium split apart to form two lighter atoms (the most common process involves a uranium atom absorbing a free **neutron** and then splitting apart into an atom of krypton, an atom of barium, and three free neutrons, which go on to split other uranium atoms). This process releases some energy as heat—although the original uranium atom was at rest, the two new atoms and the free neutrons fly away at high speed—and that heat can be used to generate electricity. Fission-based power generation has been around since the 1940s and is what most people mean when they talk about nuclear energy.

Descriptions of these processes generally say that the energy released comes from converting some of the mass of the input fuel into energy. While it is true that the mass of the products of either fusion or fission is less than the mass of the fuel, the nuclear processes that generate heat and electricity do not change the total number of particles making up the fuel. What changes is just the *arrangement* of those particles. The energy released comes not from turning material particles into energy but from the energy involved in holding the atoms together in the first place.

A dog who is used to thinking of atoms as fixed and unchanging things may not think there's any energy involved in holding the atom together—they're things that obviously just exist. A little thought about the composition of atoms, though, quickly reveals that something must be keeping them together and that huge amounts of energy must be involved. It's not hard to explain how the electrons are held in place—they're attracted to the positively charged nucleus of the atom—but the nucleus is a mystery.

---

* There's a joke in science circles that commercial electricity generation from nuclear fusion is only ten years off and is expected to remain so for the next twenty years.

The nucleus of a typical atom contains several protons packed into a ball about 0.000000000000002m across, but those protons ought to repel one another due to their positive charge. For them to be packed so tightly together requires the action of another force, one that is much more powerful than the electromagnetic repulsion of the protons but that only acts between protons and neutrons that are very close together. This strong nuclear force keeps the nucleus of the atom together and accounts for the energy that is released in nuclear reactions.

We can understand the energy release in fusion or fission by thinking about the strong force in terms of packs of dogs. The fusion process that powers the sun and other stars takes four protons (hydrogen atoms without their electrons) and sticks them together to make a helium nucleus.* The energy of the four particles stuck together is slightly smaller than the energy of the four independent particles, which might seem counterintuitive given that the two protons in the helium nucleus repel each other. Once they get close enough, though, the strong force binding them together lowers the energy. The four are happier together than alone, like four dogs forming a pack. Alone, each of the dogs must expend a good deal of energy obtaining food and so on, but with four dogs working together, they can more efficiently track prey and take care of their pet humans, so the energy expended by each dog can be lower.

For small atoms at least, the energy involved in keeping heavier nuclei together is less than the energy involved in keeping light nuclei together. If you get two light nuclei close enough together, they will fuse into a heavier nucleus and release energy in the process—about 25 million electron volts** (MeV) for the hydrogen fusion that powers the sun. The energy released is small, but the sun has huge numbers of these reactions taking place all the time, raising the temperature at its center to around 10 million degrees.

---

* In the process, two of the protons are turned into neutrons through the weak nuclear interaction. We'll talk more about the weak interaction in Chapter 12.

** One electron volt (eV) is a unit of energy equal to 0.00000000000000000016J. One million electron volts, or 1 MeV, is about twice the rest energy of an electron.

Just as adding dogs to a pack will eventually bring you to the maximum number of dogs that can work as a cohesive unit, though, the energy-lowering effect only works up to a point. As you add more particles to the nucleus, the energy gain from fusion gets smaller and smaller, until you reach iron, whose nucleus contains twenty-six protons and thirty or so neutrons. For heavier elements, the energy to hold together a larger nucleus increases, in the same way that adding another dog to a big pack can be more trouble than it's worth.

For extremely heavy elements like uranium 235, with 92 protons and 143 neutrons, the strong force can barely hold everything together. A very slight disturbance—the absorption of one more neutron—pushes it past the breaking point, and the nucleus splits in two,* like a dog pack that gets too big to manage and breaks into two separate packs. The energy involved in keeping those smaller groups together is less than the energy involved in holding the uranium nucleus together, so the fission process releases energy as heat—about 200 MeV per fission. If the reaction proceeds in a controlled fashion, this heat can be harnessed to generate electricity; if it happens in an uncontrolled chain reaction, it produces the enormous destructive force of nuclear weapons.

In both cases, the energy released comes from the strong force. When very light atoms fuse or very heavy atoms break apart, less energy is needed to hold the products of the reaction together than was needed for the fuel, and the extra energy is released as heat. While the interaction involved is different, the fundamental process is no different from what happens when an excited atom (in a fluorescent lightbulb, say) drops down in energy and releases a photon of light: the configuration of the particles making up the atom changes from a high-energy state to a lower-energy state, and the extra energy is released in some form.

Why, then, do we talk about nuclear reactions converting mass into energy when we don't use that language to describe the emission of visible

---

* Typically into a krypton nucleus with thirty-six protons and fifty-six neutrons and a barium nucleus with fifty-six protons and eighty-five neutrons, plus a few free neutrons.

light by atoms? The difference is in the scale of the energy released. When an electron in an excited atom moves to a lower-energy state or a molecule is rearranged in a chemical reaction, the energy change is around 1 electron volt (eV). While the loss of energy does change the mass, this change is completely insignificant—about 0.0000001 percent of the mass of a hydrogen atom, thus far too small to detect. Nuclear reactions, on the other hand, involve millions of electron volts' worth of energy. While this still constitutes a small change in the total mass—the 25 MeV released in fusion is about 0.7 percent of the mass of the helium created, and the 200 MeV released in fission is 0.09 percent of the mass of the original uranium atom—it is detectable. Ultimately, though, the energy released comes from interactions between particles, not the destruction of particles—the total number of particles at the end of a fusion or fission reaction is the same as at the beginning of the process.

"Isn't that kind of cheesy, dude?"

"What do you mean?"

"Well, if it's just reshuffling the particles into a lower-energy configuration, that's not *really* creating energy from mass. It's just trying to piggyback on the coolness of Einstein and relativity."

"Yes and no. The point of $E = mc^2$ is that all forms of mass and all forms of energy are equivalent. In fact, most of *your* mass comes from the same strong force that is involved in these nuclear reactions."

"What do you mean? I thought you said my mass comes from **quarks**?"

"I said that the protons and neutrons that make up the atoms in your body are themselves made up of quarks, but the quark masses are a small fraction of the total mass of a proton or neutron. A proton consists of two up quarks and one down quark, but their masses only come to 9.6 MeV of energy, while the total mass of the proton is 938 MeV."

"So what's the rest of it?"

"Well, like the protons and neutrons in the nucleus, the quarks inside the proton are held together by the strong nuclear force. The energy involved in that interaction accounts for 99 percent of the proton mass. And 99 percent of your mass comes from the same sort of interaction energy that gets released in nuclear reactions, which means that while no particles are destroyed in the release of nuclear energy, it's perfectly legitimate to say that the energy comes from the mass of the fuel."

"Which brings up an interesting question: how does all this particle stuff work? I mean, what's the deal with quarks, anyway?"

"Well, particle physics is kind of a big subject, but we can talk about it a little. After all, it's the best demonstration of the equivalence of matter and energy going in the other direction."

"What do you mean?"

"Starting with energy and using it to create particles."

"You can do that?"

"Absolutely. It happens in nature all the time, and particle physicists have raised the conversion of energy into matter to a high art. That's kind of a big subject, though, so let's give it its own chapter, OK?"

"OK, I guess. I might need some more kibble to maintain my energy, though . . ."

Chapter 8

LOOKING FOR THE BACON BOSON: $E = MC^2$ AND PARTICLE PHYSICS

I'M GRADING EXAM PAPERS at the dining room table when Emmy trots in. "Hey, dude," she says. "Where do we keep the superconducting wire?"

I'm not really paying attention, so I start to answer before I understand the question. "Hmm? Wire is in the basement, next to the—wait, what?"

"The superconducting wire. Where do we keep it?"

"We don't have any superconducting wire. And you're a dog. What do you need superconducting wire for, anyway?"

"I'm building a particle collider! I need superconducting wire for the beam-steering magnets."

"Again, you are a *dog*. Why are you building a superconducting particle accelerator?"

"Well, I've heard all this cool stuff about the Large Hadron Collider (LHC) over in Europe and how they're using it to make all sorts of new particles. And I thought to myself, 'That's a great idea. I should make one of those.' See, my food comes as particles of kibble, and I figure if I slam them together hard enough, I should be able to create whole new flavors of particles!" She's wagging her tail and drooling on the rug.

"Really."

"Yeah. I might even be able to discover the elusive bacon boson. It's responsible for making other kinds of particles yummy."

"The bacon boson?"

"It's been predicted to exist for years, but it's never been observed by any dog. It'd be the most dramatic discovery in canine physics since . . . since . . . since, like, ever."

"There's no such thing as a bacon boson."

"You're only saying that because nobody has ever observed one. But once I make my Superconducting Kibble Collider, I'll be able to find it, and then I'll be famous."

"OK, look, that's not going to work. Particle accelerators do make new particles by converting kinetic energy into mass, but it takes an incredible amount of energy to do that, way more energy than we can get around here."

"No it doesn't."

"Yes, it does."

"No it doesn't. Look, the peak energy of the proton beams at the LHC now is around 7 trillion electron volts (TeV), which is only, like, 0.00000121J. That's about the kinetic energy of a mosquito. If you dropped a 1g piece of kibble off the kitchen counter, it would have ten thousand times the kinetic energy of a proton in the LHC. And $E = mc^2$, so with that much energy, we can make all kinds of new particles." She wags her tail, looking smug.

"Falling kibble has way more energy than an accelerated proton, true, but it still doesn't amount to much. Even if you could convert all of that kinetic energy into new mass, you'd only gain about 1.1×10^{-19} kg. That's maybe 10 million atoms worth of extra kibble, which wouldn't change the flavor of anything, even if it was all in the form of bacon bosons—which don't exist."

"But they create way more stuff than that at the LHC, don't they?"

"They do, but to get 7 TeV of kinetic energy into a single proton, it needs to move at 99.99999 percent of the speed of light—which is hard to do with protons, let alone chunks of kibble."

"Oh."

"And even when they do have the proton beams cranked all the way up, they don't manage to convert all that energy into mass every time out. They're lucky to get a tenth of that—maybe one collision in 400 million produces new particles with a mass energy equal to 5 percent of the energy of the collision. You'd need to run through an awful lot of kibble before you got any bacon bosons. If they existed. Which they don't."

"Oh." Her ears droop. "How much kibble would I need?"

"If each piece was 1g, you would need something like . . . twenty thousand of the 40 lb. bags I buy for you. At $30 a bag, that'd be about $600,000."

"But—"

"And don't say that we could write a grant proposal for that. The National Science Foundation is not going to buy you kibble."

"Oh. OK."

"So, forget about building a kibble accelerator to look for the bacon boson. It's not going to work."

"All right."

"Let me hear you say it."

"I won't be building a particle accelerator in the backyard to look for the bacon boson."

"Thank you. You're a very good dog." I scratch behind her ears, then resume grading papers.

"I guess I'll have to go with my original plan, then."

"Which was?"

"Building a particle accelerator in the backyard to look for the steak quark."

No other branch of physics has managed to capture and hold the general public's interest quite as effectively as particle physics, to the point where most people tend to think that all physicists are particle physicists.* Dozens of popular books and television programs attempt to explain the current state of particle physics to nonphysicists and offer speculative theories about what might come next.

While the emphasis on particle physics can seem a little excessive at times, the field has done a lot to earn its prominent place in the public imagination. The current best theory of fundamental physics, the **Standard Model** is phenomenally successful, combining quantum physics with special relativity to describe all the known particles in our universe and their interactions. In this chapter, we take a brief look at this "theory of almost everything"** and how it explains the objects we see around us. First, though, we need to look at what it says about the particles we *don't* see and how we know they're there.

WHAT YOU SEE IS ONE-SIXTH OF WHAT YOU GET: SUBATOMIC PARTICLES

All of the material objects that dogs and humans interact with on a daily basis are made up of just three types of particles: electrons, protons, and neutrons. These combine in more than 112 ways to form the various elements of the periodic table (see the next page),† which in turn combine to form the chemical compounds making up the world we live in.

---

* Which isn't true—particle physicists account for a bit less than 10 percent of the membership of the American Physical Society. The largest single category of physicists comprises those working in "condensed matter," studying the properties of atoms and electrons in liquid and solid systems such as semiconductors.

** Borrowing the title of Robert Oerter's excellent book on the Standard Model.

From the standpoint of physics, though, protons and neutrons are not truly fundamental particles, as experiments in the 1960s and 1970s showed that protons and neutrons are made up of smaller particles called quarks. The proton consists of two up quarks and a down quark, while the neutron consists of two downs and an up. These two quarks plus the electron are sufficient to explain all the ordinary matter we see, and by adding a fourth particle, the electron neutrino,†† we can explain nearly all the nuclear reactions—the fusion of hydrogen into helium that powers the sun and other stars, the fission of heavy nuclei that is the key to nuclear power and nuclear weapons, and the transmutation of unstable elements due to radioactive decay—that we see in the ordinary universe.

According to the Standard Model of particle physics, though, these four particles represent only one-sixth of the material particle types in the universe. In addition to the two quarks that make up everything we see, there are four more types of quark and four more **leptons** (as particles like the electron and electron neutrino are called). These particles are far heavier than their everyday counterparts and are unstable, decaying rapidly into particles of the more common types. Each of the six quarks and six leptons also has an **antimatter** equivalent, a particle with the same mass but the opposite electric charge. When an antimatter particle comes into contact with its ordinary-matter equivalent, the two annihilate, converting their mass into energy; given the high abundance of ordinary matter around us, an antimatter particle usually doesn't last very long before it annihilates.

---

† The International Union of Pure and Applied Chemistry has officially recognized and named 112 elements. Elements 113 through 118 have been reported as detected in experiments but not officially confirmed as elements. Uranium, element ninety-two, is the heaviest naturally occurring element; "transuranic" elements are synthesized through a variety of different processes and last only a short time before decaying.

†† Neutrinos are, as you might guess from the name, neutral particles. They are also extremely light and amazingly difficult to detect because they interact so weakly with other forms of matter—at any instant, there are more than 40 billion neutrinos passing through every square centimeter of Earth's surface, including people and dogs. The neutrino was first postulated as an act of desperation: Austrian physicist Wolfgang Pauli found that he was unable to explain what happens to all the energy and momentum when the nucleus of an atom decays by turning one of its neutrons into a proton and spitting out an electron in the process. To account for the missing momentum and energy, he proposed in 1930 that it is carried off by a third, undetected particle with very small mass. The existence of the neutrino was eventually confirmed in 1956 by Clyde Cowan and Frederick Reines.

Some of these particles were predicted by physicists before they were detected—most famously, the positron, the antimatter equivalent of an electron, which was predicted by Paul Dirac* in 1928 and discovered in 1932. Others came as a complete surprise—when the American physicist I. I. Rabi learned of the discovery of the muon, a lepton with two hundred times the mass of an electron, he asked, "Who ordered *that*?" As is often the case with physics, the universe is a more complicated place, with many more types of things in it, than what we see around us.

"Wait a minute. What's the point of having all these extra particles, if they don't stick around long enough to make stuff?"

"Well, some of them have technological applications. Antimatter, for example, gets used all the time in medicine. It's the basis for a PET scan."

"How do you use antimatter to find out if people have animal companions?"

"Not a pet scan, a PET scan. P-E-T stands for 'positron-emission tomography' and uses small amounts of a radioactive element as a tracer to monitor how various chemicals are used in the body. When the element decays, it spits out a positron, which annihilates with a nearby electron. The scanner detects the gamma rays produced in the annihilation, and by looking at how many gamma rays come from different parts of the body, it measures the rate at which chemicals containing the tracer element are used in different areas."

"And this tells you what?"

"Well, it's one of the methods doctors use to measure brain functions or to detect specific types of cancer. It even helps in testing new types of

---

* In addition to predicting the existence of antimatter, Dirac is famous for being one of the oddest individuals in a field known for its curious characters. He was so quiet that his colleagues jokingly invented a unit in his honor, for "the smallest imaginable number of words that someone with the power of speech could utter in company—an average of one word an hour, a 'Dirac.'" Dirac is the subject of an excellent biography by Graham Farmelo (source of the above quote) titled *The Strangest Man* (New York: Basic Books, 2011) after a famous comment by the great Danish physicist Niels Bohr that of all the strange people ever to visit his institute in Copenhagen, Dirac was the strangest.

medication, to see if they get used up in organs other than the ones they're meant to treat."

"Do they use this in animal trials?"

"Yeah, sometimes."

"In that case, would it be a pet PET scan?"

"Ha, ha."

"Thank you. I'll be here all week. That doesn't answer my question, though. I mean, a positron is weird, but it's at least the antimatter version of an ordinary particle. What's the point of having the other, heavier particles?"

"Well, you might as well ask why we have squirrels and birds when we already have bunnies."

"What? That doesn't make any sense. Squirrels and birds make prey for me to chase, same as the bunnies. But these extra particles don't make stuff, so what are they for?"

"Well, nobody really knows the answer to that. We don't have a fundamental theory of everything that explains why we have all the particle types that we do and not some others. They're presumably important to the working of the universe in some way, even if we don't know what that is yet."

"Important how?"

"Well, for example, the amount of stuff produced in the Big Bang, as well as the type of stuff produced, depends on what types of particles can exist. If we didn't have all these extra particle types, we might not have all the stuff that we do in the universe. We could've ended up with a completely different distribution of particles or maybe no matter at all."

"How would you end up with nothing at all?"

"Well, the simplest models of physics predict that the Big Bang should've produced equal amounts of matter and antimatter. If that had been the case, all of the antimatter would've annihilated with all of the matter, leaving only a bunch of photons."

"That would be bad. You can't chase bunnies made of photons."

"Right. Since we see matter everywhere we look in the universe, but no antimatter, we know that there must've been more matter than antimatter created in the Big Bang. After all of the antimatter annihilated, there was a small amount of matter left over, which went on to form everything that we see."

"And the extra matter has something to do with these heavier particles?"

"Probably. We know that you don't get an excess of matter over antimatter if you only have up and down quarks and electrons, so the heavier quarks must play a role. That's not the whole story, though, because we can't explain the different amounts using just the particles we know about. So something else must be going on as well."

"Like what?"

"If I knew that, I'd have a Nobel Prize. There are a lot of theories out there, but nothing solid yet. All we can say for sure is that we know there's more matter than antimatter and that we know the heavier particle types exist."

"Which brings up an interesting point . . ."

"Yes?"

"How *do* we know all these extra particles exist? If they're unstable and decay into ordinary matter, how do we know about them in the first place?"

"Well, that's why we're talking about them in this chapter."

"Why? Oh, wait, $E = mc^2$! You make them out of energy!"

"Exactly."

MAKING STUFF OUT OF NOTHING AT ALL: PARTICLE CREATION BY COSMIC RAYS AND ACCELERATORS

The heavier exotic particles of the Standard Model are unstable, and all of them decay very quickly into particles of the more familiar types—the muon has the longest lifetime, lasting for about 0.000002 seconds before decaying into an electron and a couple of neutrinos—so we don't find them lying around in large numbers, waiting to be studied. They can be created, however, through the equivalence of matter and energy. If you accelerate a particle of ordinary matter to very high speed, then slam it into another particle, you can convert the energy of its motion into mass in the form of particles that weren't there before the collision. Every exotic particle in the Standard Model is known to physics because it is created in collisions between fast-moving particles of ordinary matter. The relationship between energy and mass is so fundamental that physicists express particle masses in terms of the energy required to create them—the electron, for example, is said to have a mass of 511 keV/c^2, because it would require 511,000 eV of energy to create one.*

The lightest exotic particles were discovered through the study of cosmic rays, with two of the first, the positron and the muon, discovered by American physicist Carl Anderson at Caltech (in 1932 and 1936, respectively). Cosmic rays are fast-moving particles of ordinary matter, mostly protons, that bombard the Earth from outer space. These particles arrive in vast numbers—thousands of them strike every square meter of the Earth's atmosphere every second—and cover a huge range of energies, with the most energetic cosmic rays ever detected packing the energy of a major-league fastball into a single proton. When these particles collide with atoms and molecules in the atmosphere, they produce all manner of exotic particles, so the simplest way to detect new particles is just to place a detector in a place that gets a lot of cosmic rays—the top of a mountain, say—and wait to see what shows up.

In the 1950s and 1960s, most of the activity in particle physics shifted to particle accelerators, and since the 1970s, all the new particle discoveries have been accelerator based. The shift was not a matter of energy—cosmic rays have been detected with energies a billion times greater than would be needed to produce any known particle—but a question of predictability. When studying particle physics using cosmic rays, you are at nature's mercy: all you can do is set up your detector and hope that a passing cosmic ray will happen to create whatever you're looking for. Particle accelerators, on the other hand, allow you to know when and where collisions will take place and approximately how much energy the colliding particles will have.

While the phrase "particle accelerator" conjures images of vast laboratories costing billions of dollars, most humans and dogs are quite familiar with the simplest sort of particle accelerator: a television set. Prior to the

* Strictly speaking, when particles are created in collisions, they appear in particle-antiparticle pairs, so you need 1,022,000 eV to make one electron and one positron.

development of plasma and LCD flat-screen TVs, the dominant television technology was the cathode ray tube (CRT), which is nothing but a particle accelerator for electrons. A CRT produces electrons by heating a piece of wire and places that wire between two metal plates with a few hundred volts between them. The hot wire is placed close to the negative plate, so electrons leaving the wire rush away from that plate and toward the positive plate, picking up speed as they go. Some electrons pass through a small hole in the positive plate and continue on to strike the inside of the screen, where they make a bright dot. This dot is moved around the screen very rapidly, using electromagnets to bend the path of the electron beam, making a pattern of dots that the human eye perceives as a picture. This is why curious human children can distort the picture on an older TV with magnets (the magnets deflect the electrons from their intended paths) and why an older TV will occasionally shock a curious dog who sniffs the screen (the electrons hitting the screen cause a slight static charge to build up and can create a spark, in the same way that a human shuffling across carpet accumulates charge and generates sparks).

The simplest particle accelerators work on the same principle as a CRT, just scaled up. A television CRT uses several hundred to several thousand volts between the plates; a small modern particle accelerator uses a million volts or more.

Voltages beyond about a million volts are difficult to create and manipulate safely, so more modern accelerators use more sophisticated techniques to accelerate bunches of particles—for example, the application of carefully timed oscillating voltages, which are easier to create and control. Another common technique is to use magnets to bend the fast-moving particles into a ring, then send two beams of particles around the ring in opposite directions. The ring technique both doubles the energy of the collision (since both particles are moving) and gives you multiple chances to get collisions with a given bunch of fast-moving particles.* Decades of

---

* The vast majority of the particles in a typical accelerator don't hit anything. The Large Hadron Collider at CERN, the world's biggest accelerator by far, sends two bunches of about 100 trillion protons each toward each other at 99.9999991 percent of the speed of light. When two bunches cross, around 20 of the 200 trillion protons actually collide; the rest of them don't hit anything at all and come back around the ring for another attempt.

slow technological improvements have pushed particle accelerators to higher and higher energies, with the Large Hadron Collider at CERN holding the all-time record of 7 TeV.

"So, there are particles out there with masses of trillions of electron volts? Wow."

"There may be particles with masses on that scale, but we haven't detected any yet. The heaviest known particles have masses in the 100 billion electron volt (GeV) range."

"They do? Isn't the LHC kind of excessive, then?"

"What do you mean?"

"Well, you need a few hundred GeV to produce a top quark and an antitop quark from energy, right? But your accelerator collides particles at 7,000 GeV. Isn't that way more than you need?"

"It's more total energy than you absolutely need, but total energy isn't the only thing involved. Just because two colliding particles have 7,000 GeV of energy doesn't mean that you'll get anywhere near 7,000 GeV/c^2 of new particles out. It's actually pretty unlikely."

"How unlikely?"

"Well, the LHC produces about 800 million collisions between protons every second. Only about two of those collisions produce a top and antitop quark with a combined mass of about 350 GeV/c^2. So it's something like a 1 in 400 million chance that you get even 5 percent of the energy you put in out as new particles."

"That's pretty unlikely. Why is that?"

"Well, because collisions of real particles are complicated. You can get a sense of this by thinking about playing pool. If you hit the cue ball straight into a stationary ball, the cue ball will stop, and the other ball will take off at the speed the cue ball had coming in. That's a collision where all of the energy of the incoming particle got transferred to the target, right?"

"Of course."

"That's not the only kind of collision you get even in pool, though. If you hit a cut shot, striking the target ball at an angle, the cue ball continues moving after the collision. That's a collision where only some of the cue ball's energy got transferred to the target. And there are a lot more ways to do that than there are ways to hit targets so the cue ball comes to a perfect stop."

"Okay. So the problem is that you keep running the protons into each other at a glancing angle?"

"That's not it exactly, but it gives you the basic idea. And the situation for a collider using protons is even more complicated, because each proton consists of three quarks. So, in order to get absolutely all of the energy out, not only would you need to get the two protons to hit exactly head-on, but you would also need each of the quarks inside the two protons to hit head-on too. Since we can't see the protons, let alone inside them, to aim precisely, it's basically just a matter of luck."

"So, what, you just throw lots of them together and hope a few will get you what you need?"

"Pretty much. You can avoid some of the problem by colliding particles that are really fundamental and not made up of anything else. The previous collider at CERN used electrons and positrons, and when those collide, you really do get all of the collision energy out as new stuff. That experiment was sensitive enough that it picked up small changes in the energy of the beam due to the effect of the tides."

"That's not that impressive, dude. I mean, all that water moving around is a big effect."

"CERN is in Switzerland, outside Geneva. It's a land-locked country. The nearest ocean is almost two hundred miles away."

"Oh. Yeah, I guess that is impressive. So why did they stop doing that?"

"One reason is money: making a higher-energy electron-positron collider would've required digging a whole new tunnel for the beam and cost a fortune. The LHC uses the same tunnel as the previous experiment but picks up extra energy by using protons."

"It's always about money, isn't it?"

"Also, even with electron-positron collisions, there are lots of different forms the energy could take, and some are more likely than others. If you want to create enough exotic particles to get a good idea of their properties, you need an awful lot of collisions, and it's easier to get protons than positrons. The LHC not only offers higher energy but provides a higher rate of collisions than you could get with antimatter, and that makes a difference. Even if you could coax electrons and positrons up to a higher energy, it would take longer to get enough collisions to be able to get a good measurement of the more exotic things."

"So, what kinds of things can you make, anyway?"

"Well . . ."

SUBATOMIC BOTANY: THE STANDARD MODEL

Particle physics can appear extremely intimidating from the outside, thanks to the dizzying array of particle types. There are so many named particles running around that when Italian physicist Enrico Fermi was corrected by a student after mixing up a name, he replied, "Young man, if I could remember the names of these particles, I would have been a botanist." The confusing lists of baryons and mesons, however, emerge from a surprisingly simple set of particles, summarized now in the Standard Model of particle physics.

According to the Standard Model, there are twelve types of material particles—six quarks and six leptons—plus their associated antiparticles. The best-known lepton is the electron, but it has two heavier cousins: the muon, whose decay helped us introduce length contraction, with about two hundred times the mass of an electron, and the tau lepton, with about thirty-five hundred times the mass of an electron. All three have the same negative charge and have positively charged antimatter equivalents. The remaining three leptons are the electron, muon, and tau neutrinos. Until relatively recently, neutrinos were thought to be massless, but experiments in the 1990s showed that their masses are merely extremely small, probably less than $2 \text{ eV}/c^2$.

Quarks also come in six varieties, which combine to form most of the particle types that confused Fermi and make particle physics so daunting for dogs. Quarks are never seen in isolation but are always found stuck together in groups of either two (making particles called mesons) or three (baryons). The lightest quarks are the up and down quarks, which combine to make protons and neutrons, as well as pions, which were first discovered in cosmic ray experiments in the 1940s.

The first sign of heavier quarks came with the discovery of the kaon, a meson discovered in 1947. The properties of the kaon were sufficiently surprising that a quantity called "strangeness" was associated with it. When the quark model was developed in the 1960s, this led the new quark to be dubbed the "strange quark." The next heaviest quark, the "up" to the strange's "down," was discovered in the mid-1970s in accelerator experiments and dubbed the "charm quark."

"See, now, those are some good names. What do you mean by 'the up to the strange's down,' though?"

"Well, the six quarks get grouped together into three 'generations,' each with two quarks. One of the two has a positive charge of two-thirds of the charge on a proton, and the other has a negative charge of one-third of the electron charge. The strange quark has a negative charge, like the down quark, while the charm quark has a positive charge, like the up quark."

"How do they figure that out, anyway? I mean, how do you measure the charge of tiny little particles?"

"Physicists determine the charge of a newly created particle by putting it in a magnetic field. Any particle with electric charge that is moving in a magnetic field experiences a force that is at right angles to its direction of motion. This force pushes a charged particle into a circular orbit, like a dog tethered to a pole in the middle of a yard, and the curvature of the path tells you about the particle's charge and mass."

"How does watching particles run in circles tell you anything about their charge and mass?"

"Well, unlike dogs, who can run in any direction they like, the direction of rotation for a charged particle depends on the charge and the field. A positively charged particle entering a detector with a magnetic field pointing up feels a force to the right. A negatively charged particle feels a force to the left."

"OK, but how does that tell you anything about the mass?"

"Well, the heavier a particle is, the more momentum it has, and the more force you need to make it turn. So, a light particle will go around in a tight circle, while a heavy particle at the same speed would make a much larger circle. A negatively charged electron entering one of these detectors makes a very tight curve to the left, while a positively charged proton with almost two thousand times the mass makes a much larger arc to the right. Physicists look at the tracks left by the particles and use their curvature to identify what went through the detector."

"So, if you saw something that followed a path the same size as an electron's, but to the right instead of to the left, you would say that was a positron?"

"Exactly. We'll make a particle physicist of you yet."

"No thanks. Unless you start colliding bunnies. That would be awesome."

The fifth quark, the "bottom," was discovered in 1977, but the sixth and final quark in the Standard Model, the "top" quark, was not observed until 1995 at Fermilab, outside Chicago. The top is far and away the heaviest known material particle, with a mass of 172 GeV/c^2, or about 180 times the mass of a proton and 330,000 times the mass of an electron. Why the mass of the top quark is so much larger than that of any of the other quarks—the next heaviest, the bottom quark, is only 4.2 GeV/c^2—is one of the mysteries of particle physics.

"Seriously, dude? 'Top' and 'bottom'?"

"What?"

"You were doing so well for a while there, but now we're back to the boring names. I mean, really, 'top' and 'bottom'? That's the best you could do?"

"Some physicists suggested the names 'truth' and 'beauty' instead."

"See, now those are some names with style. Why didn't they go with that?"

"I'm not sure. It's not my field, and the names were picked before I got into physics. I think they probably thought that 'top' and 'bottom' sounded more science-y, and that 'truth' and 'beauty' would be too hard to take seriously."

"Pathetic. That's just pathetic."

"I'm sorry you feel that way."

In addition to the six quarks and six leptons (and associated antiparticles), the Standard Model also includes twelve particles collectively known as bosons. These particles are "force carriers" that convey different interactions between pairs of particles. The electromagnetic interaction is carried by photons; the **weak nuclear interaction** is carried by the W boson (which comes in both positive and negative varieties) and the Z boson (which is neutral); and the **strong nuclear interaction** is carried by eight different varieties of **gluons.**

While it may seem strange to describe forces in terms of particles, the notion of exchange particles mediating the interactions between physical particles is one of the central ideas of modern theoretical physics, particularly the branch known as **quantum electrodynamics** (QED). This is most clearly described in terms of **Feynman diagrams,** small pictures describing physical processes developed as a calculational shorthand by Richard Feynman, which are nearly ubiquitous in theoretical physics.* Figure 8.1 shows a typical Feynman diagram representing the interaction between two electrons.

The basic idea is similar to that of the spacetime diagrams we have used before: time flows from bottom to top in the diagram, while the left-right direction represents position in space. The straight lines represent electrons, while the squiggly line represents a "virtual photon" that passes between them and communicates the repulsive force. The diagram as a whole tells a simple story about the interaction between these electrons: Once upon a time, there were two electrons near one another. One sent a photon to the other, and they moved apart. The end.

---

* Feynman developed his version of QED in the late 1940s. At the same time, Julian Schwinger and Sin-Itiro Tomonaga developed versions that look very different but produce identical results. Feynman's diagrams provide such an appealing conceptual picture that they're the most commonly used version, but all three are important for a full understanding of the theory.

Figure 8.1.

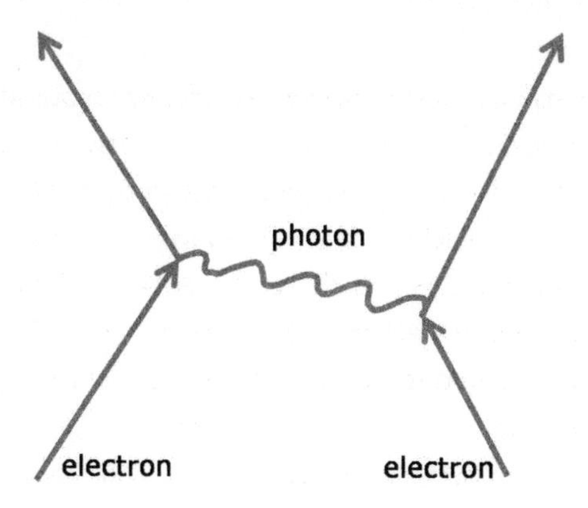

The virtual photon in this diagram acts as a messenger informing one electron of the presence of the other and telling it to move away. This not only provides a conceptual basis for understanding electromagnetism but also makes it consistent with special relativity: the photon communicating the force from one electron to another moves at the speed of light, so changes in the force take some time to be felt and do not allow faster-than-light signals. Two interacting electrons will always be in each other's light cones, and causality is safe.

The basic paradigm is one that dogs are familiar with: When one dog sees another approaching, she barks. The sound of the bark carries the message that this is her yard and the interloper should go away, in exactly the way that the photon carries the electromagnetic force from one electron to another.

"So, what, you're going to tell me that when I bark, I'm really shooting particles at other dogs? Bark-ons, or something like that?"

"Physicists do talk about sound in terms of particles, called phonons. It's critical for understanding how electrons behave in metals and superconductors, for example, but doesn't matter when you're barking at another dog. I'm just using this as an analogy for the way particles can be used to mediate interactions."

"So, how does this method handle attractive interactions? I mean, I see how this can push things apart—a really good bark drives you back a step and makes inferior dogs run away. But how does this explain attractive interactions? Or do you use some completely different method for those?"

"No, Feynman diagrams describe interactions between particles with opposite charges as well. In fact, you would use exactly the same picture as above—the space direction in these diagrams is very schematic and doesn't reflect the actual direction of motion. All that matters is that the particle changes its motion slightly."

"How do you shoot something from one particle to another and bring them closer as a result?"

"Well, when I leave you locked in the kitchen, you bark to get me to let you out, right? Barking doesn't only drive other creatures away."

"Good point."

"Anyway, you shouldn't take these too literally. It's just a sort of mathematical metaphor for the interaction."

"How is this mathematical, anyway? Pictures and stories are art, not math."

"The pictures are mathematical shorthand. Each picture represents a calculation telling you something about the effects of the interaction. When you want to calculate something, you add up the contributions of all the different diagrams that contribute, and that's your answer."

"What do you mean, 'all the diagrams that contribute'? There's just the one."

"No, there are lots, because you need to account for **virtual particles.** For example, Figure 8.2 shows the next two diagrams."

"What am I looking at here?"

"Well, on the left, one electron emits a virtual photon just before getting the photon from the other electron, then reabsorbs it a short time later. On the right, you have the same thing, but the virtual photon turns into a virtual electron-positron pair, which then disappears, turning back into a photon, which gets reabsorbed. Those are two other ways the interaction between two electrons could unfold, and you need to account for them."

Figure 8.2.

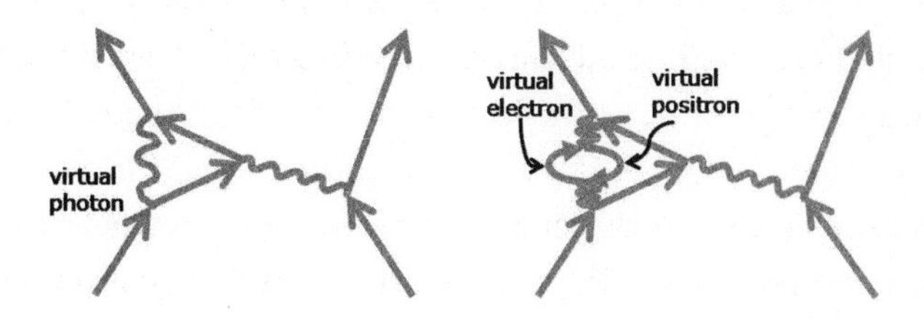

"But, wait—how did those particles just appear like that out of empty space?"

"They're a consequence of quantum mechanics, the other great theory of twentieth-century physics, which says that every bit of matter and energy in the universe must have both particle-like and wavelike properties. Beams of light, which we know have a wavelength, are made of streams of photons, with shorter-wavelength photons having higher energy, while material particles like electrons also have a wavelength that gets shorter as the energy increases. The wavelike nature means that there is always some residual energy present, even in empty space, and the particle nature means that this **vacuum energy** will occasionally manifest as particle-antiparticle pairs that appear out of nowhere and disappear again very quickly. The effects of these particles have to be taken into account, and Feynman diagrams provide a systematic way to do that."

"And you seriously expect me to believe this?"

"Explaining all the details would take a good chunk of another book,* but the predictions of QED, including all those virtual particles, have been tested and agree with experiments to something like fourteen decimal places. It's arguably the best-tested theory in the history of science."

"OK, I guess. So, particles attract or repel each other by shooting photons back and forth, like in *Star Trek*."

"It's not just photons, though. The other exchange particles—the W and Z bosons and the gluons—work in more or less the same way. They

* For example, *How to Teach Physics to Your Dog* (New York: Scribner, 2009) or Feynman's own *QED: The Strange Theory of Light and Matter* (my copy is from Princeton: Princeton University Press, 2006).

give rise to different forces, but the exchange paradigm is the same for all of them."

"And all of these things are part of the Standard Model?"

"Exactly."

The discovery in 1983 of the W and Z bosons, which had been predicted by theory, helped put the "standard" in the Standard Model, establishing it as the dominant theory of fundamental physics. Steven Weinberg, Abdus Salam, and Sheldon Glashow shared the 1979 Nobel Prize for predicting the W and Z, and Carlo Rubbia and Simon van der Meer shared the 1984 Nobel Prize for their discovery. The nature of the strong nuclear force makes it impossible to detect individual gluons, but the theory predicting their existence agrees well with experiments, and David Gross, David Politzer, and Frank Wilczek shared the 2004 Nobel Prize for working it out. The twenty-four known particles and four fundamental forces of the Standard Model represent our best current understanding of how the universe is put together.

"Wait—aren't you missing one?"

"One what?"

"Particle. You never talked about the higgy thingy."

"The **Higgs boson**?"

"Yeah, that. Every news story about the LHC mentions it, but you didn't. What's up with that?"

"The Higgs boson is another boson whose interaction with the other particles is believed to be responsible for the masses we see. It was predicted to exist by a bunch of different people, though Peter Higgs's name is the one associated with it, and we're fairly certain that it has to exist, but nobody has seen it yet. Since it hasn't been detected yet, I didn't include it in the catalogue of known particles."

"If it's all around us giving mass to things, why hasn't it been detected yet?"

"Well, because the processes that create it are very unlikely, and we haven't managed to accumulate enough collisions with the right total energy to be able to see it yet. That's one of the problems the LHC is supposed to solve—it will produce more collisions, and produce higher-energy collisions, than any previous accelerator, and if the Higgs is out there, the LHC is bound to see it."*

"And then what? Do they just close up the accelerators and go home, because physics is done?"

"Hardly. The Standard Model is the best theory we have, but we know it has some gaps. There are lots of theories of physics beyond the Standard Model, most of which predict the existence of heavier exotic particles. The LHC ought to be able to find at least some of those, which will hopefully clear up some of the outstanding mysteries in particle physics."

"But you're not sure?"

"No, but if we knew the answers in advance, we wouldn't really need to do the experiments, would we?"

"I guess. So, is that the end of the book, then?"

"Hardly. We haven't even talked about Einstein's greatest triumph yet."

"Wait, I thought that was $E = mc^2$?"

"No, $E = mc^2$ is just his most famous result. His *greatest* result, though, is the theory of general relativity, which extends the relativistic effects we've been talking about to accelerating observers and frames of reference and explains gravity in the process."

"OK, that sounds pretty cool. Let's hear it."

"That's the next chapter."

---

* They may even find it before this book makes it into print, but as of November 2011, it remains as elusive as ever.

Chapter 9

EVERYTHING IS RELATIVE IN THE MAGIC CLOSET: THE EQUIVALENCE PRINCIPLE

EMMY MARCHES UP TO MY COMPUTER as I'm checking my morning e-mail. "What the heck is the deal with relativity?!"

"Well, good morning to you too. How are you today?"

"I'm fine, but I'm confused about relativity." Sarcasm is totally lost on her.

"What are you confused about?"

"Well, you've got special relativity, right, and also general relativity. Special relativity is about clocks that run slow when you're moving and bunnies that get smaller when you chase them, and general relativity is about bowling balls on rubber sheets."

"That's just an analogy for the way that mass distorts spacetime in general relativity. There are no real rubber sheets."

"Fine, it's about bending space and bending light and black holes. The point is, it's nothing at all like special relativity. How can these possibly be part of the same theory?"

"Well, I admit, they do look pretty different—"

"That's what I just said!"

"—but ultimately, they're both about the movement of objects. The usual example people use to explain the similarities is an elevator."

"An elevator?" She looks puzzled.

"You know, like when we go over to campus? We take the elevator up to my office."

Recognition dawns. "Oh! The magic closet!"

Now I'm the confused one. "Magic closet?"

"You know. It's a tiny little room, and you go inside and close the door, and when the door opens again, you're in a different place."

"Ah. Yes, well, the point is, an elevator is basically a sealed box that you get into, which may or may not be moving. Relativity is all about what you can and can't learn about the motion from inside the elevator."

"OK, that doesn't help at all."

"Well, as I explained before, special relativity tells us that there is no absolute frame of reference, and only the relative motion between frames matters. All of the laws of physics behave exactly the same in an elevator that is moving at constant velocity as in an elevator that's standing still. There's no experiment you can do from inside the elevator to know whether you're moving or standing still."

"Like that time when we got in the magic closet, and you forgot to push any button at all, so we just sat there like idiots?" She wags her tail with exaggerated innocence.

"Yeah. Like that." We'll see how many treats she gets tonight. "Anyway, general relativity extends that to accelerating frames and gravity."

"But you said that you could detect accelerating motion. Now you're changing the story?"

"You can distinguish between accelerating motion and motion at constant speed, that's still true. General relativity adds that you can't distinguish between accelerating motion and gravity. There's no detectable difference between being in a stationary elevator in a gravitational field and being in an accelerating elevator in the absence of gravity."

"So, wait, gravity doesn't work in the magic closet?"

"No, no—gravity works. It's just that the sensation of an elevator accelerating is identical to a gravitational force. You know how when the elevator first starts up, you feel heavy for a second?"

"Yeah, I think so."

"That's the effect of acceleration, and what you feel is just like gravity getting stronger for a second."

"It's not much of an effect."

"That's because our building is old, and the elevators barely work. If the elevator accelerated faster, it would be a bigger effect. If you had a fast enough elevator, the effect could be just as big as the gravitational attraction of the Earth, making it seem like your weight had doubled. If you had an elevator that could accelerate that fast, there would be no way to tell whether you were out in space accelerating or sitting near the surface of the Earth standing still."

"That would be a mean trick. Don't do that." She looks concerned.

"It's just a thought experiment to illustrate the idea of the equivalence between acceleration and gravitation."

"Well, OK. But it's still a mean *gedankenexperiment*. Anyway, I don't see how this gets us to warped rubber sheets."

"Well, let's think about what happens to a beam of light shining across the elevator. If the elevator is standing still or moving at constant speed, the light will pass straight across the elevator."

"Right. Light always moves in straight lines."

"Not quite. If the elevator is accelerating, the light will appear to bend, as the floor catches up to it. The elevator would need to be accelerating very rapidly for it to be noticeable, but if it were, you would see that after one tick of a very fast clock, the floor had moved up by one unit of distance, after two ticks, by four units, after three ticks, by nine units, and so on, catching up to the vertical position of the light. To a person inside the elevator, it would seem like the light was following a bent path, like an object accelerating down."

"But wouldn't that tell you that you were moving? Doesn't that violate relativity?"

"It lets you know that you're not either standing still or moving with constant velocity, all right. What you can't do in general relativity is tell the difference between accelerating motion and the effect of gravity."

"But . . . doesn't that mean that light needs to bend due to gravity?"

"Exactly. For relativity to work, light needs to bend in a gravitational field by exactly the amount that it would appear to bend if we were accelerating upward at the acceleration of gravity."

"So why don't I notice light bending toward the ground?"

"The effect is really small near the surface of the Earth, because Earth doesn't have that much mass. If we look at really massive objects, though, we can see the effect of light bending due to gravity. One of the great proofs of general relativity was the observation that starlight passing close to the sun bends very slightly. Astronomers also see **gravitational lensing** where light bends around huge objects like galaxies."

"Yeah, but special relativity is all about clocks, not just light."

"General relativity also predicts that gravity causes clocks to run slow near massive objects, which we see when comparing atomic clocks at different altitudes. Both theories tell us that motion through space affects motion through time. They're really very similar."

"I guess so . . ." she says. She looks thoughtful for a minute, then picks her head up sharply. "Wait a minute. You didn't mention rubber sheets at all. What happened to bending space?"

"Ah. Well, you remember how you said that light always travels in straight lines?"

"Yeah. You said I was wrong."

"It's not exactly correct, but it's close. The correct statement is that light always travels along the shortest path between two points."

She looks really confused. "What's the difference? I mean, a straight line is the shortest distance between two points, right?"

"Not necessarily." I look around the office and pick up a basketball. "Look, imagine this ball is a globe. We're here, in Schenectady, on the *S* in 'Spalding,' and Paris is here, a bit above the *g*. What's the shortest path between those two points?"

"A straight line."

"You might think that, but the shortest path between those two points is actually what's called a great circle route, which looks curved on an east-west/north-south grid of latitude and longitude. The shortest path from here to Paris is curved."

"No it isn't. The shortest path is a straight line, through the ball."

"Well, yeah, OK, if you include all three dimensions. In reality, though, we're confined to moving on or near the surface of the Earth."

"Maybe you are. I'm a good digger. I can make a tunnel."

"Not to Paris, you can't. And we've been over this—only bad dogs dig tunnels."

"Oh. Yeah." She looks sad. "I'm a good dog, right?"

"You are an excellent dog." She wags her tail happily. "The point is, the shortest path between two points is only a straight line if you're moving on a flat surface. If you're moving along a curved surface, the shortest path between two points can be a curve, not a line."

"So . . . since light always takes the shortest path, and light bends in gravity, that means that gravity bends space?"

"You got it. It's probably better to say it the other way around—that is, gravity bends space, and light takes the shortest path, so light bends in a gravitational field—but that's how the world works. Because gravity and acceleration are indistinguishable, if you want to describe the effects of gravity mathematically, you have to describe space as curved by the presence of mass."

"Hence the rubber sheets."

"Exactly."

"The rubber sheets are a stupid example, anyway. Who has rubber sheets?"

"Well, there are plastic sheets in SteelyKid's crib. That's sort of similar."

"Maybe." She looks skeptical. "Though, I guess she does bend space and time . . ."

"From your tone, I assume you mean something other than her mass perturbing the local spacetime in keeping with general relativity."

"Well, yeah," she says. "I mean, she's not even two, and she's got you completely wrapped around her finger."

I really can't argue with that, so I don't even try.

While $E = mc^2$ is undoubtedly Einstein's most famous equation, to most physicists it (and the rest of the special theory of relativity) is merely a warm-up act for his greatest achievement, the general theory of relativity. General relativity not only encompasses all the results we have already discussed but extends them to accelerating systems (where the speed of the observer changes in time) and from there explains the force of gravity as a warping of spacetime. General relativity is a monumental accomplishment, one of the two great theories of modern physics, and the remainder of this book talks about how it works and what it means.

Many of the predictions of general relativity have become staples of science fiction (such as black holes, the subject of Chapter 10), but as fanciful as it may seem, general relativity is a rigorous scientific theory and has been tested to extraordinary precision. Countless observations from physics experiments, astronomical observations, and even everyday technologies like the **Global Positioning System** (GPS) confirm the predictions of Einstein's theory, establishing it as part of the bedrock of modern science. General relativity not only provides exquisitely detailed predictions of current physics phenomena but also provides the framework for describing the origin, evolution, and eventual fate of the entire universe (which we discuss in Chapter 11).

In this chapter, we discuss the surprisingly simple principle that gives rise to general relativity and how it leads to an even more radical revision of our concepts of space and time than their merging into spacetime. We also describe a few of general relativity's most interesting predictions and the experiments that have confirmed them to astonishing precision. Before we can do all that, though, we need to talk a bit about the most inescapable force known to dogs or humans: gravity.

WHAT GOES UP HAS ALWAYS COME DOWN: THE HISTORY OF GRAVITY

Gravity, like the air we breathe, is so ubiquitous that many early scientists didn't even attempt to explain it in detail. Prior to the early 1600s, most "natural philosophers" followed the lead of Aristotle, who held that objects fall to earth because it is in the nature of all things to rest on the ground. The weights of different objects, in this view of the world, reflect differences in how strongly they are drawn to the ground; heavier objects should thus fall faster than light ones.

The first major challenge to the Aristotelian view of gravity came in the early 1600s from Galileo Galilei, who showed that contrary to the Aristotelian model, all objects fall at the same rate, regardless of their weight.* Galileo's experiments were the first quantitative measurements of gravity and launched physics as an experimental science. Galileo's discovery also led Einstein to a crucial insight into the nature of gravity some three centuries later.

The first mathematical model of gravity comes from Isaac Newton, who proposed a universal law of gravitation describing the force between two objects.** In Newton's theory, every object with mass attracts every other object with mass, with a force proportional to the product of the

* This probably did not involve dropping weights from the top of the Leaning Tower of Pisa, as legend has it, but rather a set of careful measurements of the time required for objects to roll down ramps.

** The popular account is that Newton came up with his theory upon being hit by an apple falling from a tree in his family's orchard. While this story is probably an embellishment, it is true that Newton came up with many important ideas when an outbreak of plague closed the universities, forcing him to spend time in the country. Thus, in an indirect way, physics originates in the Black Death, which is either ironic or appropriate, depending on how you feel about the subject.

two masses and inversely proportional to the square of the distance between them. Every dog in the universe is attracted to every other object in the universe—the Earth, the sun, other dogs, and even cats—simply by virtue of her mass. If you double the mass of one of the objects—replacing a 20 kg German shepherd mix with a 40 kg Labrador retriever, say—you double the force. If you double the distance between them—moving one across the street, say—the force is reduced to one-fourth its original value.

We don't notice our gravitational attraction to objects other than the Earth because the force of gravity is extremely weak. The gravitational force between two 20 kg dogs sniffing each other from a distance of 1m is a paltry 0.00000003N, compared to the 98N force due to the gravitational force pulling each dog down to the Earth. The force is measurable, though, given a sufficiently precise technique. In 1798, British physicist Henry Cavendish, building on the ideas of geologist John Michell, made the first laboratory measurement of the force of gravity, using a "torsion pendulum" consisting of a long wooden rod suspended by a thin wire at the midpoint of the rod. Cavendish attached lead weights to the ends of the rod, essentially making a barbell. He then placed much larger masses near the ends, at carefully measured distances. The gravitational attraction between the barbell and the larger weights caused the pendulum to twist by some amount, depending on the force. Cavendish measured the twist for different masses and separations and verified Newton's predictions. Such measurements are notoriously finicky—air currents produce twisting forces larger than the gravitational force, so Cavendish enclosed the entire apparatus in a large shed and made his measurements by looking through the window with a telescope—but Cavendish's measurement of the strength of gravity was within 1 percent of the best modern value.*

Newton's theory of gravity was undeniably successful—long before Cavendish made his measurement, the theory was widely accepted because it predicts the motion of the planets in the solar system extremely well—

---

* Many different techniques have been used to measure the force of gravity over the years, but the very best measurements to date use the same torsion pendulum concept as Cavendish's original measurement.

but it has one major flaw. Newton's law of gravity predicts the force between two separated objects but provides no hint of the mechanism communicating the force from one to the other. You can use Newton's law of gravitation to determine the force of the sun on the Earth from the mass of each and the distance between them, but it does not tell you how the Earth knows about the presence of the sun 150 million kilometers away.

Newton himself was aware of this problem and famously disavowed any attempt at explanation in an afterword to the second edition of his *Principia Mathematica*:

> I have not as yet been able to discover the reason for these properties of gravity from phenomena, and I do not feign hypotheses. For whatever is not deduced from the phenomena must be called a hypothesis; and hypotheses, whether metaphysical or physical, or based on occult qualities, or mechanical, have no place in experimental philosophy.

For almost two hundred years, this was the last word on the nature of the gravitational force.

"Sheesh. That's a little snippy, isn't it?"

"It probably sounded a little less snotty in the original Latin. But only a little. It's really a response to some people who had questioned his theory on the grounds that it didn't provide a mechanism for the action of gravity. Other competing theories at the time explained gravity as the result of vortices in the aether, for example. From that point of view, Newton's theory was lacking."

"But he was right, wasn't he?"

"Absolutely. Newton's theory correctly predicts all the orbits of the planets in the solar system, which none of the other theories could manage. So he was a little annoyed to be challenged when the other theories were obviously wrong. He didn't like being questioned by those he regarded as his inferiors, which was pretty much everybody else."

"He sounds kind of like a cat, doesn't he?"

"I suspect even cats could get pointers on haughty attitude from Sir Isaac."

"Let's not tell them, then, OK?"

While Newton's theory was amazingly successful, the lack of a mechanism for gravity remained a nagging problem for physics. For a time, gravity had company, when two new forces were discovered that also mysteriously acted between separated objects: electricity and magnetism. An explanation for electromagnetic forces was eventually uncovered, though, through the idea of electric and magnetic fields permeating all of space. These were first proposed by Michael Faraday and eventually formalized in Maxwell's equations, as discussed in Chapter 2. Maxwell's equations not only provided a mathematical foundation for Faraday's field concept* but also explained the mechanism by which those fields are communicated to distant points: disturbances in the field create electromagnetic waves that travel at the speed of light. In modern terms, the electromagnetic force is carried by photons, as we saw last chapter.

Even after Maxwell's equations explained the electromagnetic field, though, gravity remained mysterious. Newton's equation for the force works brilliantly but demands an instantaneous change in the gravitational force even over astronomical distances, which was very troubling to most physicists. Nobody had a better alternative, though, so that is how things stood when Einstein took up the problem in 1907.

"THE HAPPIEST THOUGHT OF MY LIFE": THE EQUIVALENCE PRINCIPLE

The origin of general relativity is surprisingly simple and can be traced to what Einstein called "the '*glücklichste Gedanke meines Lebens*,' the happiest thought of my life," which happened sometime in 1907, only two years after the publication of his first papers on special relativity. As he described it,

---

* Faraday was a brilliant experimentalist but not very mathematically skilled, so he had trouble getting his ideas taken seriously. It didn't help that he was from the lower classes (he started his career as an apprentice bookbinder) at a time when science was mostly pursued by wealthy aristocrats.

> I was sitting in a chair in the patent office at Bern when all of a sudden a thought occurred to me: "If a person falls freely he will not feel his own weight." I was startled. This simple thought made a deep impression on me. It impelled me toward a theory of gravitation.*

Einstein had already been working on extending relativity from the special case of observers moving at constant speed to the more general case of accelerating observers, but his insight in 1907 showed that his theory was much more than a simple tweak of his earlier theory. It not only allowed him to understand the measurements of accelerating observers but provided the mechanism for gravity that had been missing since Newton's day.

The key to understanding Einstein's happiest thought is Galileo's result from three hundred years earlier: that all objects, regardless of their mass, fall at exactly the same rate. This means that to an observer in free fall, accelerating toward Earth at the 9.8 m/s^2 common to all falling objects, there appears to be no gravity. Every other free-falling object in the vicinity falls at the same rate and thus appears to be at rest. If there were any difference between the effects of gravity and any other force, some objects would fall faster than others and allow the falling observer to distinguish between falling and floating in space at rest. Without such a difference—and no such difference has ever been detected despite countless tests**—Einstein realized that his principle of relativity applies to observers in free fall: a dog falling under the influence of gravity is perfectly entitled to view herself as being at rest.

"That's just crazy! Of course you can tell the difference between falling and floating at rest: when you're falling you see yourself rushing toward the ground."

---

* Both this quote and the one in the paragraph above are taken from Abraham Pais's outstanding scientific biography of Einstein, *Subtle Is the Lord: The Science and the Life of Albert Einstein* (New York: Oxford University Press, 1982).

** The most spectacular demonstration came when Commander David Scott of the Apollo 15 mission dropped a hammer and a feather on the surface of the moon; in the absence of air, the light feather and heavy hammer hit the ground at the same time.

"Right, but that's an inference based on experience, like back in Chapter 1. Mathematically, though, there's no difference between accelerating downward and being at rest while everything else accelerates upward. This is why most discussions of relativity talk about observers in sealed boxes, like an elevator that may or may not be falling."

"Yeah, but when you're falling, you feel all funny in the stomach."

"Right, you feel weightless, because your internal organs are falling at the same rate as the rest of you and no longer have their weight supported by the rest of your body. So the cases that are perfectly equivalent are being in an elevator in free fall—because the cable broke or some such—and an elevator floating in space, far away from any other objects. Out in space, there isn't any gravity, while in the free-falling elevator it just seems like there's no gravity."

"Yeah, but what's an elevator doing out in space?"

"I don't know. It's a thought experiment. Presumably, it's some sort of Adventure, like Winnie-the-Pooh might get involved in. Floating in Space while Awaiting Rescue. That sort of thing."

"Quoting obscure science fiction novels* is not an explanation, dude."

"Yes, but it amuses me. Also, it doesn't matter. Just roll with it, OK?"

"All right. What about a stationary elevator, though? When it's not moving, I feel gravity just fine."

"That's the other key insight. When the elevator is standing still, objects inside it fall to the ground at a constant acceleration. But you can get exactly the same effect by accelerating a space-going elevator in the 'upward' direction: when you release an object inside the elevator, it no longer experiences a force to keep it accelerating with the rest of the elevator. In keeping with Newton's laws, it continues along through space in a straight line at constant speed, while the floor of the elevator accelerates up to it. To an observer inside the elevator, though, it looks just like the object fell when it was dropped."

---

* Daniel Keys Moran's *The Long Run* (New York: Bantam Spectra, 1989) is a tremendously enjoyable but hard to find book. An electronic version is available from Moran's website at http://danielkeysmoran.blogspot.com.

"How's the elevator accelerating in space?"

"I don't know. It's a thought experiment. It's a rocket-powered elevator or something."

"Maybe that's how they got way out in space in the first place?"

"Sure, if that makes you happy. The important thing is that there's no way to distinguish between the effects of gravity and the effects of any other acceleration."

"So, falling down an elevator shaft is the same as floating in space?"

"Yes."

"And sitting in a stationary elevator is the same as accelerating upward in space?"

"Yes."

"And this is the key insight that explains gravity."

"That's the starting point, yes."

"I don't know, dude. I don't think I'm willing to just take your word on this one. I'm going to need more than that."

"Well, keep reading . . ."

The observation that all objects, regardless of mass, fall at the same rate is the simplest form of the **equivalence principle.** It says that the mass used to determine the force of gravity from Newton's law of gravitation is the same as that used to determine the acceleration caused by a force through Newton's second law of motion. As a matter of pure mathematics, there's no reason why those two masses need to be the same, but countless experiments have shown that, in fact, they are.

Einstein's great realization in 1907 was that he could extend this equivalence principle to encompass all of physics, in the same way that his principle of relativity extended Galilean relativity to include Maxwell's equations and all the rest of physics. He realized that the equivalence was not mere coincidence but an indication of a deeper equivalence that affects all of physics.

Just as the core of special relativity is a strengthening of the Galilean principle of relativity, the core of general relativity is a strengthening of the existing equivalence principle. Einstein's version of the equivalence principle has two parts:

1. Physics according to an observer who is free-falling due to gravity is indistinguishable from physics according to an observer who is stationary in an inertial frame without gravity.
2. Physics according to an observer who is stationary in a gravitational field is indistinguishable from physics according to an observer who is accelerating in a frame without gravity.

These two parts are easy to demonstrate with the motion of ordinary objects, as Figure 9.1 shows. Emmy the dog in a frame without gravity would see her favorite squeaky ball float motionless in the air near her, but the same effect can be obtained in a free-falling elevator. Inside the elevator, Emmy sees the ball hang in midair, not falling, but Nero the cat outside the elevator says that this is just because both dog and ball are falling at the same rate.

Figure 9.1.

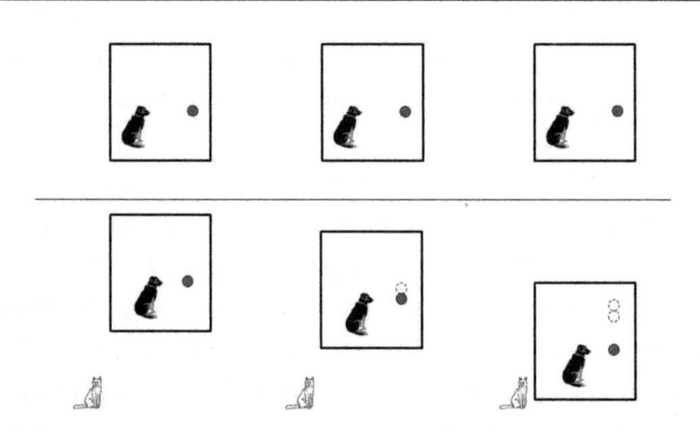

Similarly, a stationary dog in a frame with gravity sees a dropped ball fall with a constant acceleration, but the same effect can be obtained in an elevator without gravity accelerating in the upward direction. A cat floating in space outside the elevator says that the apparent fall is really the floor of the elevator accelerating upward to meet the ball (see Figure 9.2).

Figure 9.2.

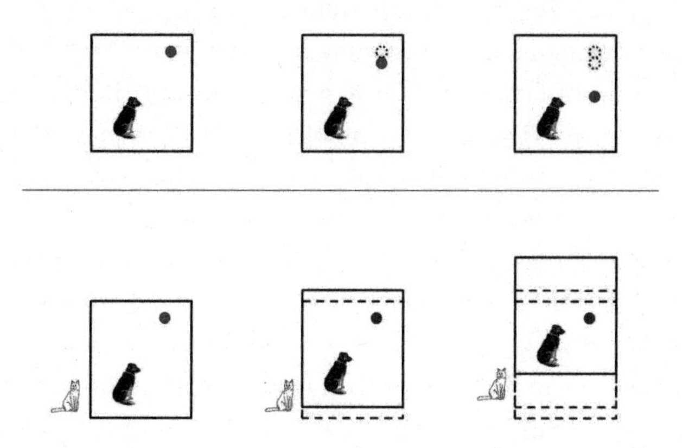

Whether in the falling or the accelerating elevator, Emmy will believe that she is in an inertial frame, where Newton's laws hold, at least as long as she can't see the world outside the elevator. Within the elevator at least, all the predictions of special relativity should hold, which means that all of the laws of physics work the same for Emmy inside the elevator and Nero outside.

Emmy and Nero will disagree about the details of what happens and why, just as a dog and a cat moving at constant speed relative to one another disagree about the details of time and length measurements. Just as special relativity provides a mathematical and conceptual framework for reconciling the different observations of cats and dogs moving at constant speed, general relativity provides a mathematical and conceptual framework for reconciling the different observations of accelerating cats and dogs or cats and dogs at different places in a gravitational field. And just as special relativity requires rethinking the fundamental nature of space and time, general relativity requires further rethinking of spacetime, as well as a fundamental revision of the nature of gravity.

Although Einstein's first glimpse of this theory came in 1907, he didn't publish the final version of general relativity until 1915. It is this, not $E = mc^2$, that physicists regard as his greatest contribution to science.

"Wait, if the idea is so simple, why did it take him eight years to publish it? What was he doing, carving it in stone?"

"Well, for one thing, Einstein moved around a bunch. His 1905 papers got him a series of faculty positions, first in Zurich, then Prague, then Zurich again, then Vienna, and finally Berlin. More than that, though, general relativity is really complicated. The central idea is simple, but expressing it mathematically is extremely difficult. Einstein basically had to reinvent a whole bunch of mathematics before he could get it right."

"So, wait, he moved five times *and* invented new branches of mathematics? Wow, he *was* a genius!"

"He did move five times, but I said *re*invent mathematics. The math had been worked out earlier, most notably by German mathematician Georg Friedrich Bernhard Riemann in the nineteenth century."

"So, wait, if it's old math, then what took Einstein so long?"

"It was known to mathematicians, but Einstein's background was in physics, and physicists didn't use the math in question at that time. So he had to learn it for himself, and independently reinvent parts of it, with the help of his friend Marcel Grossmann."

"Oh. I thought physics was all about math?"

"Physics is a highly mathematical science, but it doesn't use *all* of mathematics. There have always been problems that were interesting to mathematicians but not to physicists—or, at least, problems that physicists didn't know were interesting to them."

"You mean this happens a lot?"

"Oh, yeah. In addition to Einstein's recreating **Riemannian geometry** for general relativity, Werner Heisenberg laboriously reinvented matrix math while working on a theory of quantum physics. And Murray Gell-Mann spent a bunch of time reinventing parts of group theory when he came up with the quark model of matter. Physicists are always stumbling across new techniques, only to have mathematicians say, 'Oh, that. We knew about that fifty years ago.'"

"You probably ought to talk to them more, then."

"Yeah, we're working on that."

EVEN LIGHT FEELS GRAVITY: PREDICTIONS FROM THE EQUIVALENCE PRINCIPLE

The strong form of the equivalence principle put forth by Einstein lets us predict and understand new phenomena by moving back and forth between an observer in free fall and an observer in a gravitational field. According to Emmy inside the elevator, there is no gravity, and all the normal laws of physics apply. Nero, watching the elevator fall under the influence of gravity, sees the elevator accelerating and thus interprets the events inside the elevator differently. Nero's observations, though, must be a combination of the physical laws seen by Emmy and the effects of acceleration. The difference between Nero's view from outside the elevator and Emmy's view from inside tell us how gravity affects the physics being considered.

This recipe—describe a simple scene from Emmy's point of view, then Nero's, then compare the two to find the effect of gravity—lets us make three striking predictions: light bends due to gravity, vertical beams of light vertically change frequency, and clocks run slow in a gravitational field. All three of these have been experimentally tested to great precision, making general relativity one of the best-tested theories in the history of science.

The first key prediction concerns the bending of light in a gravitational field. Let's imagine that Emmy has a laser pointer inside the elevator with her and shines it horizontally across the elevator onto the opposite wall. She sees the light beam move across the elevator in a perfectly horizontal line, with a pulse of light leaving the laser at some height above the floor of the elevator and hitting the opposite wall at exactly the same height above the floor.

What does Nero see in this case? Both Nero and Emmy *must* agree that the spot on the far wall of the elevator is the same distance above the floor of the elevator as the laser. They won't necessarily agree about what that distance is—remember our earlier discussion of length contraction—but they'll agree about the relative height. According to Nero, though, the elevator has moved some distance during the time it took the light to travel across it, so he sees the light hit the wall at a position lower than where the light was emitted by the distance that the elevator fell during this time.

This looks similar to the light clock argument of Chapter 3, with one important difference. In the light clock thought experiment, an observer moving at constant speed relative to the clock sees the light take a longer path but still follow straight lines. In this case, the elevator is accelerating, so the light follows a curved path. If you doubled the width of the elevator, Nero would see the elevator fall four times as far as in the smaller elevator, and thus the light must fall four times as far. Nero sees the light beam moving across the elevator follow the same sort of parabolic trajectory as a tennis ball thrown out for a game of fetch, as seen in Figure 9.3. And since any difference between Nero's view and Emmy's is due to the influence of gravity, this tells us that light bends due to gravity.

Figure 9.3.

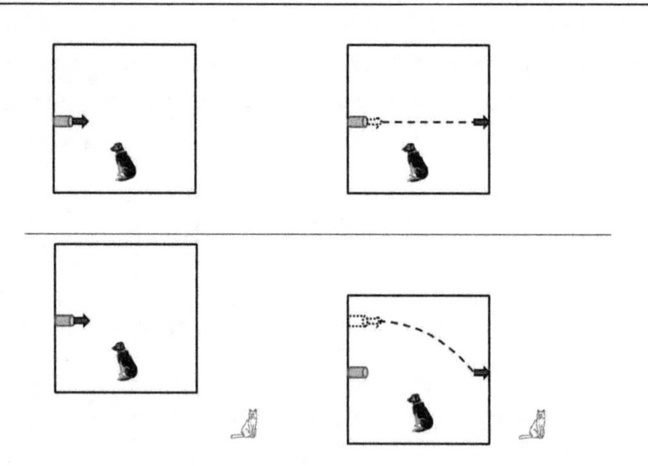

The bending of light in a gravitational field is a dramatic prediction, but it is extremely difficult to observe because the speed of light is so great. Light crosses an elevator with a width of 6 ft. in about 6 ns, during which time the elevator falls about 0.0000000000000002m, approximately one-fifth the diameter of the nucleus of an atom. There's almost no hope of detecting this bending on Earth.

The bending of light becomes observable, though, if we beef up the source of gravity. The gravitational pull of Earth is too small to have much effect, but a vastly larger object, like the sun, with the mass of almost a million Earths, bends light enough to see. The effect is still tiny, but if you look close to the rim of the sun during a solar eclipse, the bending of light shifts the apparent position of the stars behind the sun.* The effect of the sun's gravity is a little like the distortion caused by placing a magnifying glass over a page of text: light rays coming from the text are bent by the lens, making them seem to come from different places on the page.

This bending became the first great proof of general relativity, when a British expedition led by Arthur Eddington photographed the sun during a total eclipse in two different places: Sobral in Brazil and Príncipe Island off the coast of Africa. Eddington's team painstakingly measured the positions of stars in their photographic plates, and after months of careful work, they found that the stars were shifted by exactly the amount predicted by general relativity. The results, announced in November 1919, made Einstein an overnight celebrity. The normally stodgy *New York Times* got a little carried away, announcing the results with no fewer than six headlines:

LIGHTS ALL ASKEW IN THE HEAVENS

Men of Science More or Less Agog Over Results of Eclipse Observations

EINSTEIN THEORY TRIUMPHS

Stars Not Where They Seemed or Were Calculated to be,
but Nobody Need Worry

A BOOK FOR 12 WISE MEN

No More in All the World Could Comprehend It, Said Einstein When
His Daring Publishers Accepted It

* The shift is always there, but the stars are only visible during an eclipse.

"I love the 'nobody need worry' bit. Because I was totally going to lose sleep over the stars being a tiny bit distorted near the sun."

"Like I said, they got a little carried away."

"So, is that it? I mean, you go through all this business about light bending, but the only effect is a tiny little shift that you need to spend months looking for in pictures of an eclipse, which almost never happens?"

"No, that's not the only example, just the most famous, because it's the event that made Einstein the world's most famous scientist. The bending of light due to gravity turns up all over the place in astronomy. For example, there's the Einstein Cross" (see Figure 9.4).

"Five smudgy little blobs?"

"Each of those blobs is a galaxy. The four on the outside are all images of the same galaxy, which is actually behind the one in the middle. The gravity from the millions of stars in the middle galaxy bends light around it, producing those four images."

"How do they know it's just one galaxy seen four times and not five different galaxies that just happen to be close to one another?"

"Well, for one thing, it would be really unlikely for five galaxies to sit in exactly this pattern. More importantly, though, they can measure the spectrum of light from the four images, and they're identical. Since individual galaxies emit slightly different colors of light, the fact that these four are the same tells us that they're really one galaxy seen multiple times."

Figure 9.4. Image of the Einstein Cross, the gravitational lens G2237 + 0305 taken by the European Space Agency's Faint Object Camera on board NASA's Hubble Space Telescope. (Image not subject to copyright; taken from www.commons.wikimedia.org/wiki/File:Einstein_cross.jpg.)

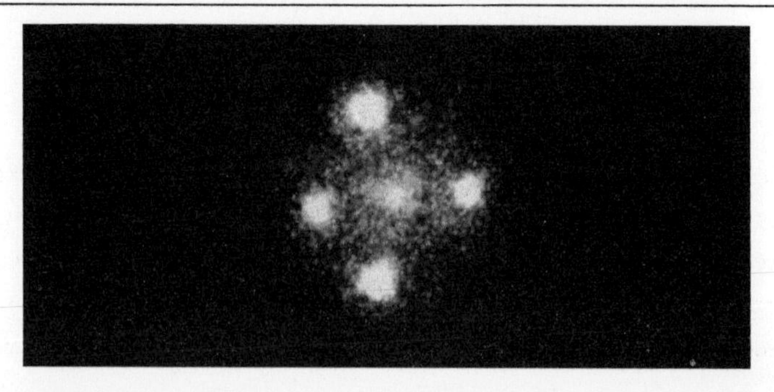

"OK, so you've got some stars moving during eclipses and four smudgy blobs. What else?"

"Astronomers use this kind of gravitational lensing to map out the distribution of matter in the universe. By looking at how light from distant objects is distorted when it passes near massive objects that are closer to us, astronomers can figure out how much mass those closer objects contain and how that mass is arranged."

"This is all very cosmic, but does it have any practical consequences? Or is it all stuff that you need giant telescopes to detect?"

"The bending of light doesn't have much practical application, I'm afraid, but it's the most famous effect of general relativity. Some of the others have important consequences, though."

"All right, then, let's move on to the next example."

DON'T IT MAKE MY RED LIGHTS BLUE: GRAVITATIONAL FREQUENCY SHIFTS

The next major phenomenon predicted by the equivalence principle can also be seen by shining light across the elevator. This time, though, we change the direction of the light beam to be vertical, along the direction of gravity. Emmy mounts a laser pointer on the ceiling of her elevator, sending a beam of light straight down, producing a spot on the floor directly below the laser (see Figure 9.5). Nero outside the elevator agrees about the direction of the light, but Nero and Emmy disagree about the *color* of the light when it strikes the floor of the elevator.

This may seem even stranger than light bending due to gravity, but we can understand the shift using the **Doppler effect,** a shift in the frequency of waves from a moving source that is most familiar when applied to sound waves. Sound produced by an approaching object—say, a car with its horn blaring, rushing toward a dog who is minding her own business on the side of the road—is detected at a higher frequency than the same sound emitted by a stationary source. Sound produced by a receding object—the same noisy car continuing past the dog down the street—is detected at a lower frequency. This produces the characteristic eeeeee-owwwwww

Figure 9.5.

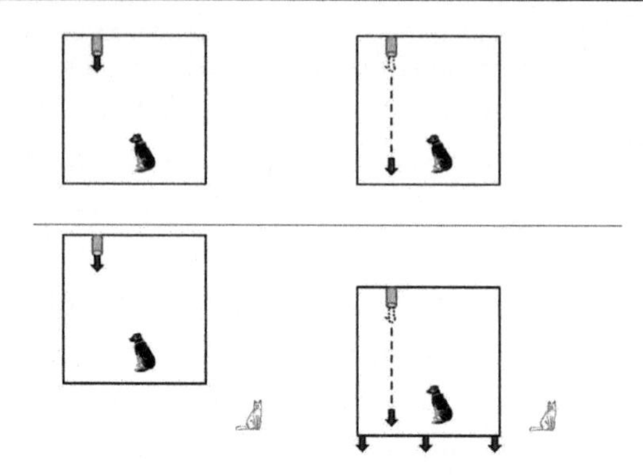

noise of a fast-moving car: as it comes toward you, the sound shifts higher in pitch, then, as it goes away, the sound gets lower in pitch, with a rapid change from high to low occurring just as the car passes you. And, like all laws of physics, relative motion is what really matters, so an observer approaching a stationary source hears it at a higher frequency, just as if the source were moving.

The Doppler effect with sound waves is familiar because we regularly encounter objects whose speeds (compared to the speed of sound) are fast enough to produce significant shifts. The Doppler effect affects light waves as well and can be used to determine the velocity of light-emitting objects: if you know the frequency at which the light is emitted (say, because you know what the object is made of and thus the characteristic colors of light it emits), you can determine its velocity relative to you by comparing the frequency you observe to the frequency you expect.

The middle line in Figure 9.6 shows the spectrum of the four colors of visible light (represented by the four peaks) emitted by a sample of hydrogen gas, according to a dog sitting at rest next to the gas. The upper line is the spectrum that would be observed by a cat moving toward the dog at 2 percent of the speed of light; it shows a characteristic blueshift, with all four peaks moving toward the blue end of the spectrum (higher frequency, shorter wavelength). The lower line is the spectrum that would be observed by a bunny running away from the dog at 2 percent of the speed of light;

Figure 9.6.

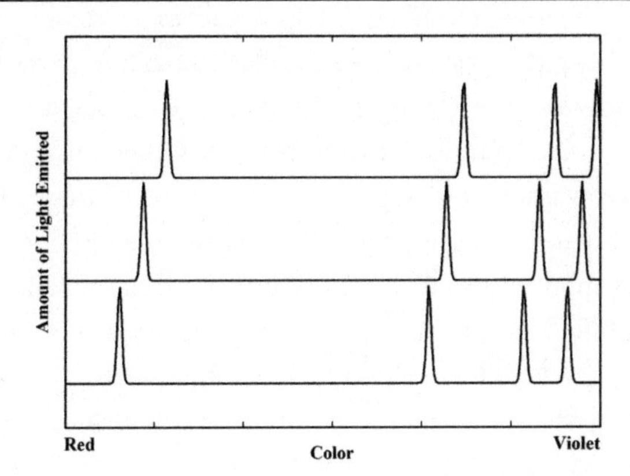

it shows a characteristic redshift, with all four peaks moving toward the red end of the spectrum (lower frequency, longer wavelength). The high speeds in this example make for dramatic shifts, but with sensitive spectrometers, physicists and astronomers can detect Doppler shifts of light due to velocities as small as a few centimeters per second, which is a standard technique for determining the velocity of distant astronomical objects.*

What does the Doppler effect have to do with gravity? Well, from Nero's point of view outside the falling elevator, the floor of the elevator is accelerating—which means that by the time the light emitted by the laser on the ceiling reaches the floor, the floor is moving away from it at a higher speed than when the light was emitted. Thus, a detector placed on the floor of the elevator should see the light redshifted by an amount depending on the acceleration of the elevator and the time required to cross it—the taller the elevator, the greater the shift.

Inside the elevator, though, Emmy doesn't see any such shift—as far as she can tell, the light striking the floor has exactly the same frequency as the light emitted by the laser. Emmy and Nero have to agree about this point—they will disagree about the exact value, but both must see Emmy's

* In fact, the technique of measuring velocity by Doppler shift is so well established that it is used to detect planets around distant stars by detecting changes of only a few meters per second caused by the gravitational tug on the star as the planet moves around it.

detector record the same frequency as was emitted—so Nero must see the frequency of the light shift up as it moves from the ceiling to the floor, in a way that exactly compensates for the Doppler shift from the floor moving away. Emmy sees the light emitted and detected at the same frequency, while Nero sees the light reaching the detector shifted up in frequency as it moves downward, and he says that Emmy sees the frequency shifted back down due to the Doppler effect from the moving floor. A similar chain of reasoning shows that light sent upward in the falling elevator must be redshifted as it moves up, compensating for a blueshift from the ceiling moving toward the laser with increasing speed.

Since the effects of the falling elevator must be indistinguishable from the effects of gravity, this tells us that light sent from higher to lower elevation must shift up in frequency, while light sent from lower to higher elevation must shift down in frequency. This **gravitational redshift** has been verified experimentally, with the first test done in 1960 by Robert Pound and Glen Rebka at Harvard University. They placed samples of radioactive iron at the top and bottom of a 22.5m "tower" (really, the top and bottom of the physics laboratory building) and showed that gamma rays emitted by atoms at the top of the tower had a different frequency when they reached the bottom. Fittingly, they measured the size of the shift through the Doppler effect by showing that atoms in the detector at one end of the tower would absorb light emitted by identical atoms at the other only if they moved the detector up or down at the appropriate speed. The frequency shift is tiny—corresponding to only a few millimeters per hour—but Pound and Rebka's apparatus was sensitive enough to detect the shift and show that it agreed with the predictions of general relativity.

"So, wait, if you stand at the top of the stairs, and I look up at you, you look bluer? And if I stand at the top of the stairs and look down, you look redder?"

"Exactly. The shift is extremely tiny, so you'd need an incredibly sensitive spectrometer to pick it up, but there is a real shift in the frequency of light emitted at one altitude and detected at another."

"So, is that why the sun looks red when it sets?"

"What?"

"You know, because it's lower down, closer to the horizon. So we're looking down at it, and it looks red."

"No, that's not why the sun looks red in the evening. The sun isn't really getting any lower, remember, it's just the Earth spinning."

"Oh, right. Sometimes I forget."

"The red color is because the light from the sun passes through more of the atmosphere when it's setting than when it's overhead, and the atmosphere scatters blue light more effectively than red light."

"Which is why the sky looks blue?"

"Exactly. Blue light headed past us gets scattered by the atmosphere, making the sky look blue. The sun looks red near the horizon because the longer path means that more of the blue light has been bounced away, to become the blue sky seen by people to the west of us."

"Or east, in the morning."

"That's right. There is a tiny gravitational redshift due to the sun, because light leaving the surface of the sun is effectively traveling 'up' all the way from there to here. The light reaching us from the surface of the sun is very slightly redder than the same light on Earth. It's a tiny effect, though, and very difficult to distinguish from the shifts due to the motion of gas in the sun. People spent a long time looking for it, but the Pound and Rebka experiment was the first to make a really good measurement of the gravitational redshift."

"That's pretty clever, I'll grant, but it's still arcane physics-lab stuff. Does this business have any practical application?"

"Not in the form of the gravitational frequency shift itself, but the next effect has a major consequence. It also clears up the one loose end I left hanging in the previous couple of pages."

"Would this be why I don't see the frequency shift in the falling elevator? Because that was my next question."

"That's it exactly."

"See, you can't get anything past me."

THE CURIOUS INCIDENT OF THE DOG, THE LIGHT, AND TIME: GRAVITATIONAL TIME DILATION

Although the discussion of gravitational red- and blueshifts above explains why the cat sees the shifted frequency of the light, it seems to create an entirely new problem: how can two observers who are not moving relative to one another see a change in the frequency of light emitted by one and received by the other. When we're moving between Emmy's free-falling frame and Nero's frame with gravity, the shift makes sense, but in the Pound-Rebka experiment, both the source and detector were at rest.

To put the problem in more concrete terms, imagine two dogs at opposite ends of a vertical tower 100m high, each with an identical source of light and a detector that records its frequency. Both dogs measure their own source as having some frequency but measure a different frequency for the other dog's source: Maeve the springer spaniel at the top of the tower sees her source oscillate 1,000,000,000,000,000 times in one second, while light sent up the tower from the bottom oscillates only 999,999,999,999,990 times. Winthrop the basset hound at the bottom, on the other hand, says the light sent down from above oscillated 1,000,000,000,000,010 times in one second. But the two sources are identical, and neither dog is moving relative to the other, so how can this be?

To paraphrase Sherlock Holmes, the curious thing here is not what the dogs see, but what they *don't* see: If the frequency of light changes as you move up or down, why doesn't Maeve notice the shift in her own source? Why does she only see Winthrop's light shifted in frequency, not her own?

The answer is the most profound consequence of the equivalence principle, a simple but radical revision of our notion of space and time: while the dogs are stationary relative to one another, because they are at different positions relative to the Earth, their clocks run at different rates. Clocks higher up in a gravitational field tick at a faster rate than clocks at lower elevation.

The shift in the clock rate is tiny, but it's just enough to explain the observations of the two dogs in their own frames of reference. Winthrop at the bottom of the tower says that Maeve records light coming up at a lower frequency because her clock ticks too fast—what she sees as one second is a little shorter than one second on his clock, so she hasn't counted enough oscillations. And Winthrop sees light coming down with a higher frequency because the clock at the top ticks too fast, meaning that the real frequency must be higher to pack 1,000,000,000,000,000 oscillations into the shorter second of Maeve's clock.

Maeve, on the other hand, sees the reverse: she sees her clock run at just the right speed, while Winthrop's clock ticks too slowly. That explains why Winthrop records a too-high frequency for light sent down from the top and why the light sent up from the bottom has a lower frequency than it should.

In keeping with the principle of relativity, this **gravitational time dilation** applies to any imaginable clock. No matter how the dogs choose to mark the passage of time—atomic clocks, quartz watches, counting tail wags—Maeve at the top of the tower sees time pass more quickly than Winthrop at the bottom. Light oscillates more rapidly, watches tick more quickly, dogs age more swiftly—any process that depends on time happens faster at high altitude than low altitude.

The idea that gravity affects time may seem absurd, but the shift has been measured directly, using the same ultraprecise aluminum ion clocks described back in Chapter 3, where we talked about a comparison between two clocks, one based on a nearly stationary ion and one whose ion was moving at some small velocity. These clocks are so precise that they detect the shift predicted by special relativity for speeds below 10 m/s. In a second part of the experiment, scientists from the National Institute of Standards and Technology used their clocks to test general relativity as well, this time keeping both ions at rest but using hydraulic jacks to lift one of the two clocks up by 33 cm.

Figure 9.7 shows eighteen measurements of the difference between the two clocks, the first thirteen with both clocks at the same level and the last five with one clock raised 33 cm (a bit more than one foot) above the other. The shift is tiny—in one hundred seconds of operation, the light associated with the upper clock oscillates just four more times than the light associated with the lower clock (out of 100,000,000,000,000,000 total

Figure 9.7. Open circles show measurements taken with both clocks at the same height; filled circles show measurements with one clock raised 33 cm above the other. Horizontal lines represent the average of the different data sets; the dashed lines show the uncertainty. (Modified from C. Chou et al., "Optical Clocks and Relativity," *Science* 329, no. 5999 [2010]: 1630–1633; used with permission.)

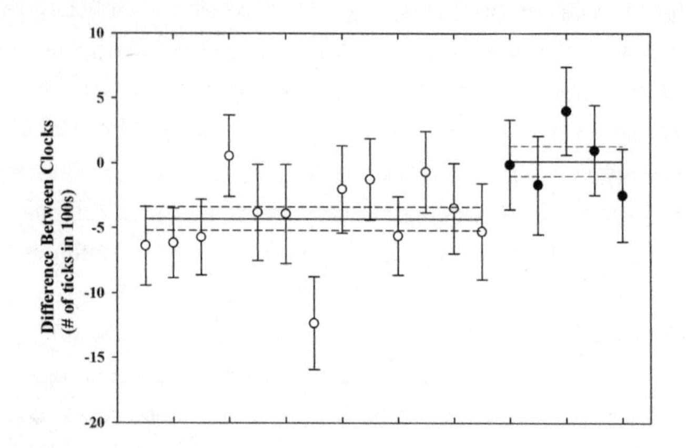

oscillations)—but the upper clock clearly ticks faster than the lower clock, exactly as predicted by general relativity.

Of course, a larger change in altitude produces a bigger shift, and it is here that general relativity plays an important role in technology. The GPS receivers that humans use to navigate are based on a network of satellites orbiting the Earth. Each satellite contains an atomic clock and broadcasts a radio signal identifying the satellite and giving the time according to that clock. A receiver on the ground picks up the signal from several satellites and calculates the distance to each satellite using the difference in the signals' arrival times. This allows the receiver to determine its position on Earth's surface to within a few meters.*

For GPS to work, the timing has to be accurate to within 10 to 20 ns, which is why they need atomic clocks—no mechanical clock could provide the necessary timing accuracy. The satellites orbit high above the Earth, though, so they are both fast moving (about 14,000 km per hour) and

* The radio signals from the satellites travel at the speed of light, which is the same for all observers, so the difference between the times received from various clocks lets you find the distance to each. The orbits of the satellites are very well known, so knowing the distance from three satellites picks out a single spot on Earth.

higher up in the Earth's gravitational field (about 20,000 km altitude). The high orbital speed means that the clocks run slow by about 7 μs per day, in accordance with special relativity, compared to a clock on the ground. Their high altitude, on the other hand, means that they run fast by about 45 μs per day, in accordance with general relativity. The net effect of these two theories is that the satellite clocks gain about 38 μs per day on identical clocks on the ground.

In 38 μs, light travels about 11 km, so without the correction, GPS measurements of positions could drift by up to 11 km per day, farther than most dogs would care to walk. The impressive accuracy of GPS navigation is only possible because the scientists and engineers who designed the system understood general relativity and deliberately adjusted the clocks in the satellites to run slow by 38 μs per day before they were launched. With this correction, the clocks in orbit tick at the same rate as clocks on the ground, keeping the whole system in synch.

"See, this comes from having tiny, defective noses. If you could smell properly, you wouldn't need these fancy technological systems to navigate. Then you wouldn't need to worry about the effects of gravity on time."

"True, but no matter how you navigate, the time dilation effect of gravity is still real. You may not need to worry about it, but it's there. And, anyway, I think you sell human noses short."

"Dude, no offense, but you couldn't smell your way out of a paper bag."

"True, but by not having so much of our brains tied up processing smells, humans free up a bunch of cognitive capacity for other things—like thinking up physics theories and developing sophisticated technological tools. Which is why *we* rule the world, and *you* do the happy dance when you get table scraps."

"I dunno. I think your brains are overrated. The opposable thumbs are the real issue—if I could work a doorknob, I would be unstoppable."

"Whatever. Anyway, are you satisfied?"

"With table scraps?"

"No, silly, about general relativity being important. The US government spent millions of dollars establishing the Global Positioning System, which only works if you take general relativity into account. Is that a practical enough application for you?"

"Millions of dollars? That's a lot of kibble. Yeah, I guess that works."

The equivalence principle is the inspiration and starting point for general relativity, but it is not the whole picture by any stretch. The phenomena we've talked about in this chapter can be understood qualitatively by thinking about the equivalence principle, but the principle by itself won't get you all of the answers—in particular, using just the equivalence principle to calculate the deflection of light by the sun gets you a prediction that is half as big as it should be. Einstein originally published the too-small value in 1911, working from an early version of the theory, and corrected his prediction after completing the full theory in 1915.*

The equivalence principle is adequate for explaining most of the more ordinary phenomena predicted by general relativity. The full theory adds the notion of curved spacetime to the equivalence principle, leading to even more exotic predictions. For those, though, we need truly enormous masses, of the sort only found in astronomical situations—giant stars, collapsed stars, and even the universe as a whole. We'll look at the full theory, curved space and all, in the next chapter.

"I was just thinking that this discussion was notably lacking in rubber sheets."

"Really?"

"Well, really, I was just thinking that it would be nice if you gave me some treats and maybe a walk, but I thought about the rubber sheet thing, too."

"I thought that might be the case. This chapter's starting to run long as it is, so it seemed like a good place to take a break."

"Good plan. And as long as we're taking a break anyway, how about some treats and a walk?"

---

* Fortunately for Einstein, though not for the scientists involved, a series of earlier expeditions to look for the bending of starlight during an eclipse based on his 1911 prediction ran into various mishaps—bad weather, broken equipment, and, most dramatically, the outbreak of World War I in 1914, which landed one German scientist in a Russian POW camp for some time—and were unable to measure any shift. Eddington's 1919 expedition was the first to successfully measure the bending of light, and their value agreed with the full and correct theory of 1915.

Chapter 10

WARPING THE UNIVERSE: GENERAL RELATIVITY AND BLACK HOLES

I'M SITTING AT THE TABLE with books spread out in front of me, putting together a lecture for tomorrow's class, when Emmy trots up behind me. "Hey, dude, what'cha doin'?" she asks.

"I'm getting my lecture together for tomorrow's class."

"I could help with that, you know."

"Writing my lecture? I doubt it. I've seen your attempts at writing. You usually end up eating the pen."

"Writing is overrated. I've got you to write things for me!"

I can't argue with that.

"Anyway, that's not what I meant. I meant I could help *teach* your class, by demonstrating relevant phenomena."

"Really."

"Yeah, I'm a great demonstrator! What are you lecturing about?"

"Conservation of momentum in elastic and inelastic collisions."

"Oh, that's easy. I could run around your class bumping into stuff. It'd be great."

"You know, I don't think my students have trouble visualizing what a collision looks like. I'd rather spend the time going over the math, if you don't mind."

"Okay. That's too simple an example, anyway. I'm much better at demonstrating more exotic phenomena. For example, I can do a great demonstration of a black hole."

"Really."

"Yeah, really. For one thing, my fur is mostly black. And you could scatter treats all around the room, and I could gobble them up, the way a black hole gobbles up anything in its path."

"Why do I suspect that the treat gobbling is the whole point of this exercise?"

"Dude." She gives me a wounded look. "I am seriously interested in educating young humans about the wonders of astrophysics."

"Sorry. Anyway, there's a big problem with your analogy."

"Yeah, what's that?"

"Well, anything that makes it past the **event horizon** of a black hole can never return to the outside universe. The food that you eat, on the other hand, manifestly does not remain within you, which is why I have to carry plastic bags with us when we go for walks."

"Ah." She thinks for a moment. "No problem dude, I have the answer."

"Really?"

"**Hawking radiation.**"

"I can't believe I'm having this conversation."

"No, I'm serious. Stephen Hawking—you know, the guy with the wheelchair and the computer-y voice—proved back in the 1970s that black holes emit radiation. Right?"

"Sure. Virtual particles are created near the event horizon of the black hole, and some of the particles will fall in and be lost. When that happens, the second particle of the pair flies off and becomes real. In order to conserve energy, the black hole loses a corresponding amount of mass. So, over time, the black hole is whittled away to nothing."

"So, the treats I eat represent the matter falling into the black hole, while—"

"I see where you're going with this. Don't even think about it."

"I guess that means you don't want to hear my take on the **black hole information paradox,** then?"

"Dear God, no. Anyway, the event horizon is really one of the least interesting features of a black hole, from a physics standpoint. And you can't demonstrate the really cool features."

"Really?"

"Really. The coolest thing about black holes is the way they warp spacetime in their vicinity. For example, time passes much more slowly near a black hole than out in space. If you made a quick orbit around the event horizon of a black hole, then returned to a point far from it, you might easily find that people watching you saw your trip take hundreds or thousands of times as long as you thought it took."*

"Yeah, well, um, I slow time because I'm so good that people can spend hours rubbing my belly."

"Space gets distorted too. Near a black hole, fundamental geometric relationships that we're used to in flat space can behave in completely different ways. You might measure the three angles in a triangle, for example, and find that they add to something other than 180 degrees. Or you might find that the circumference of a circle drawn around a black hole is only twice as big as its diameter, instead of the familiar value of pi."

"Yeah, well. . . ." She thinks hard for a minute. "Ummmm . . . I like to eat pie! Sorry, dude, I got nothing. I guess I can't demonstrate a black hole as well as I thought."

---

* This figures in the plot of Brian Greene's children's story *Icarus at the Edge of Time*.

I rub her behind the ears. "That's OK. We love you even though you don't produce significant distortions of spacetime."

"Can I still come to your class and collide with stuff?"

"No." Her tail droops. "You have much more important work to do here."

"What do you mean?"

"Well, when you're here, I feel safe leaving all our stuff in the house, because I know you're guarding it."

"That's right! I'm an important component of our home defense system!"

"Exactly. I mean, without you barking like a crazy thing every day, who knows how badly the mail carrier would've robbed us by now?"

"Thanks for reminding me. Go back to writing your lecture, dude. I have to patrol." She trots off, tail waving happily, a dog with a renewed sense of purpose.

If ordinary humans or dogs use the phrase "Einstein equation," odds are pretty good that they mean $E = mc^2$, Einstein's most famous equation. When physicists refer to the Einstein equation, though, odds are good that they mean the Einstein field equation from general relativity:

$$R_{\mu\nu} - \frac{1}{2} R g_{\mu\nu} = 8\pi G T_{\mu\nu}$$

It's not as widely known as the equivalence between mass and energy,* but to many physicists it's the only equation out there that deserves to be called *the* Einstein equation. It's a deceptively simple-looking equation—the symbols with subscripts stand not for single numbers but for tensors, which can be thought of as 4 × 4 grids of numbers, with their own special rules for multiplication and so on—and describes the way spacetime bends in response to the presence of matter.

---

* Though it does turn up in some odd places, such as the opening frames of the French animated movie *The Triplets of Belleville*.

This **curvature** of spacetime is the mathematical core of general relativity, which John Archibald Wheeler* pithily summarized as "Matter tells space how to curve, space tells matter how to move." General relativity explains gravity in terms of a mixing of space and time, like that discussed in Chapter 5, but varying from place to place. An object in free fall is not pulled by a force at a distance but is following the simplest, straightest path it can through a universe in which space and time are themselves curved.

In this chapter, we talk about the geometry of curved spacetime and how this determines the behavior of everything in the universe. We look at some examples demonstrating the effects of general relativity within our solar system. Finally, we look at the most spectacular example of general relativity in action, a black hole, an object so dense that not even light can escape its gravity.

THE WORLD WARPS ITSELF AROUND YOU: GENERAL RELATIVITY AND CURVED SPACETIME

All of the phenomena we talked about last chapter, strange as they are, come from applying the equivalence principle to some simple situations. There's more to general relativity than just the equivalence principle, though. Einstein's insight regarding the equivalence principle set him on the path toward general relativity, but there were many years of hard work ahead and a lot of mathematics to develop along the way. The full theory is highly mathematical, more so than most humans want to deal with, let alone dogs. We can get a sense of how it works, though, by putting together pieces we have already discussed.

Back in Chapter 5, we talked about the mixing of space and time in special relativity. Two events that one observer sees as separated only in time—happening in the same place—another observer may see as

---

* Coauthor, with Charles Misner and Kip Thorne, of *Gravitation* (New York: W. H. Freeman, 1973), the definitive book on the subject, which is about two appendices away from producing measurable space-time curvature itself.

separated in space as well as time, and vice versa. We can understand this mixing through a particular rule for combining distance in space and distance in time to get an invariant spacetime interval. The speed of an observer through space affects the rate at which they move through time, but the right combination of space and time lets us reconcile all the observations of different moving observers.

The equivalence principle tells us that one effect of gravity is to change the rate at which observers move through time in different *places*: clocks at higher altitude tick faster than those at lower altitude. The full theory of general relativity, then, must combine space and time in a way that lets us reconcile these observations—that is, it must be a system in which you use different rules to calculate the spacetime interval at different points in space.

A system where the measure of distance depends on where you are may seem bizarre and improbable, but in fact we use such a system all the time when we describe the location of objects on Earth's surface using latitude and longitude. A change in latitude corresponds to the same distance everywhere—about 111 km for every degree—but a change in longitude corresponds to different distances at different places, decreasing closer to the poles. To cover 180 degrees of longitude at the equator, a dog must run half the circumference of the Earth, about 20,000 km. At 43 degrees north, the latitude of Schenectady, the distance would be just under 11,000 km, and at the Arctic Circle, just 3,300 km. Right at the North Pole, a dog could run 180 degrees of longitude without getting out of breath.

"Dude, there are all kinds of problems with this. I mean, your equatorial dog would need to run through miles and miles of ocean."

"It's a hypothetical, OK? Imagine it as happening on an alien planet just the size of Earth but with no oceans, if you prefer."

"And the dog at the North Pole would have to run through Santa's workshop, dodging stupid elves."

"..."

"Just kidding!"

"Well, that's a relief."

"More seriously, though, isn't this a problem for that theorem from the German mathematician named after me?"

"Emmy Noether?"

"Right. You said momentum is conserved because the laws of physics are the same everywhere. Now you're saying that how you measure distance depends on where you're doing the measuring. Isn't that a contradiction?"

"No, because the laws of physics do work the same way everywhere. That's the whole point: every observer sees the same physical laws. The changing of space and time is the manifestation of gravity, which is one of those laws."

"But if you measure distance different ways in different places, doesn't that mess things up?"

"No, because the change depends on the presence of a massive object. It's not a change in the laws of physics any more than the presence of a magnet attracting metal objects would be. The point of Noether's theorem is that the mathematical rules describing forces between objects don't depend on where those objects are located but only on their separation. A massive object here on Earth bends spacetime in the same way as it would out in the orbit of Jupiter. The rules used to determine that bending are the same everywhere, and that's what really matters."

"Oh, OK. Speaking of rules, what are they—the rules used to determine the bending of space and time, that is?"

"I was just getting to that . . ."

The rules for understanding how spacetime is bent by a massive object are mathematically very similar to the rules for describing the geometry of objects on a curved surface. Such curved geometries behave very differently than the normal flat geometry that dogs and humans are familiar with—the most obvious consequence being that parallel lines in a curved geometry will eventually cross.* The rules of geometry in curved space

---

* For an example, think about two lines of longitude: at the equator, they are parallel in the usual sense, but if you keep extending those lines along the surface, they meet at the North and South poles. You can also have situations where parallel lines diverge, for example, lines drawn on the surface of a saddle.

were worked out in the 1800s by numerous mathematicians, most notably Bernhard Riemann. The final theory of general relativity is expressed entirely in terms of this Riemannian geometry,* most famously the Einstein field equation on page 226. The left side of this equation is all about the geometry, particularly the metric $g_{\mu\nu}$, which expresses the rules used to determine distance at a point in spacetime. The right-hand side describes the distribution of mass, energy, and momentum around that point. Most of the business of general relativity involves using this field equation to calculate the curvature caused by a particular distribution of matter.**

That equation by itself doesn't provide much illumination to many physicists, let alone dogs, though, so let's think about what it means using spacetime diagrams. In Chapter 5, we talked about the light cone containing all those events in spacetime that a particular observer can affect or be affected by while moving and communicating at speeds less than or equal to the speed of light. Every observer has a light cone for every instant in time, and however they move, their trajectory through spacetime must fall entirely within their own light cone. In the flat spacetime we have discussed previously, different observers disagree about the rate at which time passes and the exact distance in space or time between two events, but they have one thing in common: when we draw a spacetime diagram showing their worldlines, the light cones all point in the same direction.

Figure 10.1 shows the worldlines of three objects at rest in a world without gravity—say, the Earth, Nero the cat floating in space some distance from the Earth, and Emmy the dog floating in space a bit farther out—which follow straight vertical lines into the future. If we look at a series of instants—represented by two-dimensional slices through space in our picture, because it's impossible to draw four-dimensional figures on paper—

---

* In contrast to Euclidean geometry, the more familiar, flat-space version developed by the ancient Greek mathematician Euclid around 300 BC. The fact that it took 2,000 years to come up with the curved-space version tells you how difficult the problem is.

** So much so that the physicist and raconteur Richard Feynman once managed to get to a conference whose location he had forgotten by telling a cab dispatcher that he wanted to go to the same place as a bunch of guys who "would have their heads kind of in the air, and they would be talking to each other, not paying attention to where they were going, saying things to each other, like 'G-mu-nu. G-mu-nu'" (from his autobiographical *Surely You're Joking, Mr. Feynman* [New York: W. W. Norton, 1997]).

Figure 10.1.

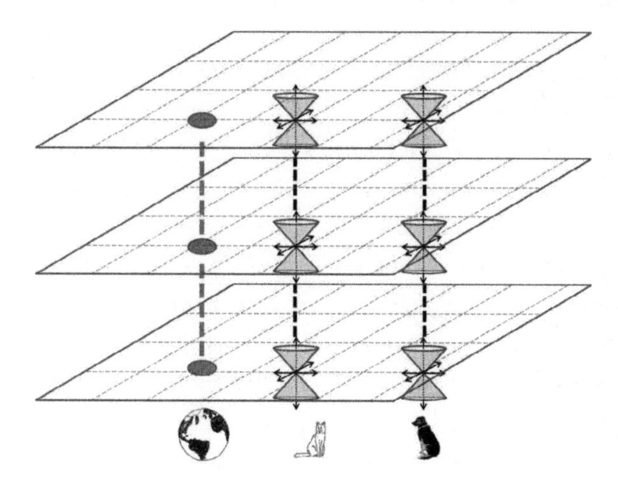

we can draw a light cone for Emmy and Nero at each instant: two 45-degree cones meeting at their points, aligned vertically along the worldlines for the two animals.

What does this diagram look like in curved spacetime—that is, if we turn gravity back on? According to general relativity, the large mass of the Earth bends spacetime in its vicinity, causing a deflection of the path of any object moving nearby. We can visualize this as a change in the orientation of our observers' light cones (greatly exaggerated in Figure 10.2 in order to make it visible). Nero, close enough to Earth to feel a significant gravitational pull, has his light cone inclined slightly toward the Earth; Emmy feels less of a pull, so her light cone remains nearly vertical.

What does this mean? Nero doing nothing, as cats tend to, always moves straight up the time axis of his light cone. But now, thanks to the bending of spacetime, that axis points a little toward the Earth. As a result, even though he isn't doing anything to cause himself to move, his natural trajectory curves toward the Earth—at the next slice through spacetime, he is closer to Earth, and his light cone is bent over even farther, so he moves toward the Earth

Figure 10.2.

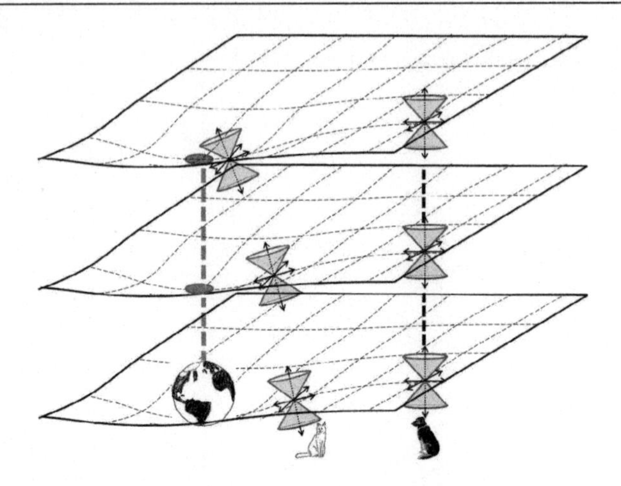

faster, and so on. Nero is in free fall, accelerating toward Earth at a speed determined by gravity. Emmy, meanwhile, is farther away; thus, her light cone is barely affected, and she follows a basically vertical path, as before.

"So, wait, the speed of light is lower for light headed away from the Earth?"

"No, it's the same for everyone."

"OK, but the cat's light cone is tipped sideways. The outward side of his light cone isn't at 45 degrees anymore. Doesn't that mean it's not moving at the speed of light anymore?"

"The light cone is tipped, but the space and time axes tip with it. From Nero's point of view, everything looks normal—the time axis goes along the center of the light cone, and the position axes go out to the sides. According to him, the speed of light is exactly the same as it was without gravity."

"But . . . they're tilted."

"This is why general relativity is so difficult to get your head around. What Nero sees as just time looks like a combination of space and time to you, and vice versa, but neither of you can make that comparison directly. All you can measure is what's going on right by you and what you receive via light signals from elsewhere."

"So, how do you get to draw this sort of picture then?"

"Well, we can infer it from the measurements made by you and Nero and the equations of general relativity."

"OK, I guess, but what do I see?"

"Well, as you're measuring things from a distance, light signals coming from Nero do indeed take longer to arrive than they should. That is, if he flashes a light on and off in rapid succession—for example, sending a Morse code signal saying, 'Help, help, I'm falling toward Earth'—you will see the light pulsing more slowly and the pulses arriving later than expected."

"Which means that the light was traveling slower!"

"That's one conclusion you might jump to. Another way to interpret it, though, is as a combination of two effects: time running slower down where he is and space being slightly stretched. That is, you would say the signals are slow because his clock is slow and because the distances he measures are a little longer. That's why he gets the same value for the speed of light—his clock ticks a little too slowly, but his rulers are wrong, too, and those effects cancel out. On the path from Nero to you, though, those effects add together, delaying the arrival of the signal."

"Isn't that kind of complicated?"

"Yes, but it's central to the theory. The slowing of his clock is just gravitational time dilation, but the stretching of space is new. If you tried to calculate the delay in a signal's arrival time from just the time dilation, you'd get a value half as big as the real answer. You have to include the stretching of space as well."

"But I thought this was all about the equivalence thingy?"

"The equivalence principle is the beginning of the theory, but to put all the mathematical pieces together, you need space to stretch too. That's what Einstein struggled with for all those years—he had the equivalence principle part by 1911, but it didn't get everything. When he worked out the full theory in 1915, he quickly found that a number of his earlier predictions, including the bending of light by the sun, were half as big as they should be. When he corrected them, though, they agreed very nicely with experiments like Arthur Eddington's eclipse observations and the well-known perihelion shift of Mercury."

"Perry who?"

"Sorry. The orbits of planets going around the sun aren't circles but ellipses, so there's a point in every orbit that is the closest that planet gets to the sun, right?"

"Sure."

"When you include the gravitational pull of the other planets, you find that the orbits don't quite make perfect ellipses, but change their orientation by a tiny bit each orbit. That closest point, the 'perihelion' position, moves around very slowly."

"OK, I guess. I assume this has been measured?"

"Yes, very precisely. In fact, it was used to predict the existence of previously unknown planets—way back in 1846, the planet Neptune was discovered because astronomers couldn't explain the observed orbit of Uranus with just the known planets. The French mathematician Urbain Jean Joseph Le Verrier realized that he could explain Uranus's orbit using another large planet orbiting further from the sun and predicted where that planet should be. When astronomers looked, sure enough, there it was."

"And this has what to do with relativity?"

"Neptune has nothing to do with relativity, but there is a similar shift in the orbit of Mercury, the closest planet to the sun, that nobody could explain. Some people even speculated that it might be caused by another planet even closer to the sun, but nobody could find one. One of the first things Einstein did when he finished general relativity was to check its prediction for the perihelion shift of Mercury, and he found that it agreed beautifully with observations. This was the first solid evidence for the theory, and Einstein later said his excitement over the discovery gave him heart palpitations."

"Physicists are so easily excitable."

"Says a dog who goes nuts chasing a laser pointer spot on the floor . . ."

"What?! Where? Oh—I see what you did there. Anyway, I still don't see what this has to do with rubber sheets."

"Well, keep reading."

The tilting of the light cone tells us that time and space at Nero's location are mixed, according to Emmy: what he sees as space, she sees as a mixture of space and time. Nero near the Earth measures different results than Emmy would guess trying to measure those same quantities from a long distance away. This tilted light cone picture also nicely explains Nero's fall toward Earth: the Earth is not just a nearby point in space but a point in his future—if he doesn't exert any effort, his worldline will inevitably intersect that of Earth.

One way of presenting all this information together is an **embedding diagram,** treating each of our two-dimensional slices through space as if it were embedded in a fictitious third dimension, with the mass of the Earth stretching spacetime into this third dimension in its immediate vicinity. The traditional example is the stretching of an elastic material like a sheet of rubber, but a more canine analogy might be a large rock dropped into a muddy yard: the yard without the rock is flat, but as the rock sinks into soft ground, it creates a depression around itself that will cause smaller objects nearby—clods of dirt, tennis balls—to roll toward it in the same way that the bending of spacetime pulls objects toward the Earth. It also increases the distance that a small creature moving along the surface—an ant crawling along the ground, say—measures between a point far from the rock and the center of the rock. The ant has to crawl down the slope to get to the rock, which means it follows a longer path than a purely horizontal trajectory would give.

In Figure 10.2, we see that our two-dimensional slices get distorted, so the distance along the path from the dog to the cat stretches out a little. This stretching is tiny but has been measured directly in the solar system—the straight-line distance from Earth to Mars is about 60 km longer* than you would expect in flat space when that line has to pass near the sun. This stretching of space has been confirmed by monitoring radio signals to and from Mars probes.

Nero's tilting light cone, meanwhile, is always perpendicular to the surface. His spatial axes point along the surface, as they must, while the time axis points up from it. This gives us a convenient way to picture what will

---

* That's out of around 370 million kilometers, so it's a very small change, but one that's measurable with modern communications technology.

be seen by any observer at any point on the surface—we can just draw a light cone perpendicular to the surface and use that to trace out what happens to that observer in free fall.

"So, the Earth has moved down in the time direction, and that's why clocks on the surface tick slow?"

"No, but that's a common source of confusion. The third dimension in these embedding diagrams is completely fictitious. It's just a visual aid to let you see how the curvature of spacetime affects measurements of distance."

"So there's no way to bounce off the rubber sheet like a trampoline? Because that was going to be my next question."

"No, there isn't. The real structure of spacetime is a four-dimensional thing that can't be represented well with a two-dimensional medium like paper. To capture all the details, what you would really need is a four-dimensional surface representing the universe, stretched and curved in a fictitious fifth dimension."

"Ow. That makes my head hurt."

"Right. That's why we stick to imagining two-dimensional surfaces. Drawing two dimensions stretched in a fictitious third dimension already taxes my drawing abilities to the limit."

"I've noticed. So, getting back to this tilty light cone business: is there a limit to how far you can tip these things?"

"Well, the amount of tilt depends on the curvature of spacetime, which in turn depends on the mass and the size of the thing producing the curvature."

"Aren't they the same thing?"

"No, because different objects have different densities. If I drop a 10 lb. bag of kibble in the yard, it won't make much of a dent, because it's spread over a large area. If I drop a 10 lb. rock, though, it sinks into the ground because the same mass is concentrated in a smaller area."

"OK, I guess that makes sense. But still, isn't there a limit to the curvature?"

"No, there isn't. If you pack enough mass into a small enough point, you can get any amount of curvature you like."

"But . . . doesn't that get weird?"

"Yes, absolutely. The result is a black hole, one of the strangest objects known to science."

"Oh, right, black holes. Talk about those next."

"As you wish."

YOU CAN CHECK OUT ANY TIME YOU LIKE BUT YOU CAN NEVER LEAVE: BLACK HOLES

Thinking about the curvature of spacetime in terms of a tilting light cone leads naturally to the question of how far you can tilt the light cone of an observer near a massive object. For example, can you tip it by 45 degrees? If so, the worldline of the outward light ray in our diagram would be perfectly vertical, not moving at all in space, according to a dog observing from a safe distance. Even heading away from the object at the speed of light will not allow this observer to escape falling into such an object.

The mathematical description of such an object was discovered within a few months of the publication of Einstein's theory of general relativity by German physicist Karl Schwarzchild, who somehow managed this while fighting on the eastern front in World War I.* Schwarzchild used Einstein's equations to show that if you compress a sphere of matter so that its radius becomes smaller than a critical value,** the resulting object curves spacetime so strongly that light from the object can never reach the outside world. A massive object smaller than this **Schwarzchild radius** becomes a black hole, an object that even light cannot escape.†

---

* Schwarzchild reported his discoveries in a December 1915 letter to Einstein that concludes, "As you see, the war treated me kindly enough, in spite of the heavy gunfire, to allow me to get away from it all and take this walk in the land of your ideas." While serving in the trenches, he contracted a rare disease and died in May 1916.

** The exact value of the radius is given by the formula $R = \frac{2GM}{c^2}$ and depends on the mass M, the gravitational constant G, and the speed of light. Any object, regardless of mass, will become a black hole if it is compressed far enough.

† The name "black hole" was introduced by John Wheeler in the 1960s and quickly caught on, despite sounding mildly dirty in French.

The idea of gravity affecting light predates relativity—back in 1783, the British physicist John Michell had used Newton's theory of gravity to develop the idea of "dark stars," bodies vastly more massive than the sun, which would also prevent light from escaping. Michell's dark stars depended on the idea, favored by Newton, of light as a stream of particles, and while it generated some interest, it was quickly dropped when experiments in the early 1800s demonstrated the wavelike nature of light.

A Schwarzchild black hole, while similar to Michell's dark star, is a much stranger object than can be produced by Newtonian physics. A dark star would've sent light outward in all directions, but the speed of that light would've decreased as it moved farther from the star, eventually stopping and turning around to fall back, like a tennis ball thrown for a dog to chase. At any distance from the star, you would see some light from it, thanks to the few photons that managed to reach that far before turning around.

In a Schwarzchild black hole, on the other hand, no light emitted from inside the Schwarzchild radius will ever reach the universe outside that radius. Light emitted from that radius would take an infinitely long time to reach an outside observer and would be redshifted to infinitely long wavelengths in the process. For this reason, the Schwarzchild radius is also called the event horizon of the black hole: it's the last point from which we can obtain any information about what's going on. After an object enters the event horizon, we can never know what happens to it.

Black holes bring out the full weirdness of general relativity, so let's look at how the predictions of general relativity play out on an imaginary mission to explore a black hole. Let's imagine two observers in a spaceship near a black hole: Emmy manning the controls of the mother ship at a safe distance from the black hole and Nero piloting a smaller probe shot into the black hole, never to return.

"Dude? Thank you."

"For what?"

"This is the *best* thought experiment *ever*!"

"Oh. You're welcome. I guess."

A long distance from the black hole, both Emmy and Nero agree about the orientation of their light cones. Both measure the same speed of light, and if Nero uses a green laser to send a pulse of light to Emmy every second, she will agree that one green pulse arrives every second. As Nero moves toward the black hole, his light cone begins to tilt relative to Emmy's. She sees the light pulses he sends shifted into the red and finds that they arrive at a slower rate.

As Nero falls toward the black hole, the spacetime curvature becomes more extreme, and his light cone tilts even further. The light pulses shift to longer and longer wavelengths, first into the infrared and then the microwave and radio regions of the spectrum. The time between pulses gets longer as well, stretching to seconds, then minutes, then hours. Nero's final signal before he crosses the horizon of the black hole takes an infinite amount of time to arrive and is shifted to an infinitely long wavelength. No matter how long Emmy waits, she will never see Nero cross the horizon but only an increasingly "red," increasingly faint image of him frozen at the moment of the crossing.

"Sucks for that cat. It must be really weird, to be in a place where everything slows down and turns red. I'm glad I got to stay on the spaceship and watch."

"You're forgetting the central message of relativity again. Nero doesn't see time slow down. He sees all the laws of physics work in exactly the same way they did on the ship."

"Yeah, but he never crosses the horizon. *Something* must be different."

"No, you never *see* him cross the horizon."

"Isn't that the same thing?"

"Not at all. Read on."

If Emmy never sees Nero cross the horizon, what does that mean for him? From Nero's point of view, most of the effects Emmy witnesses are reversed. The time between signals from her decreases, and the signals are shifted up in frequency, becoming first blue, then ultraviolet, then turning into X-rays. Unlike Emmy, though, Nero doesn't see anything special happen when he reaches the horizon—the horizon isn't a physical barrier, just a point in space. He passes right through, with nothing to mark the event.

Does this mean that everything is rosy for Nero? Hardly. Once inside the horizon, his eventual fate is sealed: his light cone has tipped over so far that there is nothing he can do to avoid falling all the way to the **singularity** at the center of the black hole, where all the hole's mass is concentrated into a single infinitesimal point. The singularity has been Nero's inevitable destination for the entire time he's been falling inward, but once he crosses the horizon, even moving at the speed of light (the maximum possible speed) in the outward direction can't save him from eventually encountering the singularity. The singularity has ceased to be a position in space and has become an inevitable event in Nero's future.

Is encountering the singularity really all that bad for Nero? As far as we know, yes. As he moves closer to the singularity, the curvature of spacetime continues to increase, and eventually tidal effects start to become important: the force on his feet* is greater than the force on his back, so he finds himself stretched.

These **tidal forces** are always present, but at large distances from the singularity, the effect is small because the tidal force reflects the change in the curvature of spacetime, which is only detectable over large distances. As you get closer to the black hole, the curvature changes more rapidly, and eventually it changes significantly over the size of the cat. The difference between the force on Nero's feet and that on his back is the tidal force, stretching him out along the direction of his fall and compressing him in the perpendicular direction.** As he gets closer and closer to the

* We'll assume his feet are pointing toward the black hole because cats always land on their feet.

** As you can probably guess, the tidal force is called that because the same sort of interaction is responsible for the tides in Earth's oceans: gravitational forces from the moon (and to a lesser extent the sun) stretch Earth along the direction between Earth and the moon and compress it in the perpendicular directions. The effect on Earth's crust is hard to see, but the water in the oceans moves relatively easily and produces the tides.

singularity, the tidal forces get stronger and stronger, until eventually he is pulled apart and stretched into a long thin string, a process some have dubbed "spaghettification."

"Dude, that's gross. Not even an arrogant cat deserves that sort of treatment."

"True. That's why it's a good idea to avoid falling into a black hole."

"Yeah, but how can you avoid it? I mean, it's got infinitely strong gravity."

"You avoid it the same as anything else. If you're well outside the horizon, there's no difference between the gravitational field of a black hole and anything else of the same mass. It's not like in the movies, where a black hole forms and the force suddenly increases a hundredfold, sucking everything in like a vacuum cleaner. If you built a hollow shell the size of the Earth around a black hole with the mass of the Earth, you could walk around on it just like you walk around on the Earth. You'd never know the difference."

"But anything that falls into it never comes back, so wouldn't it eat the hollow shell?"

"Anything that falls in never comes back, but it's not any harder to avoid falling into a black hole with the mass of the Earth than it is to avoid falling into the Earth, provided you're outside the horizon."

"So, you're telling me that when those scientists in Geneva make a black hole with the Large Hadron Collider, it's not going to eat the Earth?"

"No, it's not. For one thing, if particle collisions can create black holes, it's already happening due to the ultra-high-energy cosmic rays that we talked about in Chapter 8. Those haven't destroyed the Earth. And even if they did make a black hole, it would have a tiny mass, thus a miniscule horizon—much smaller than the size of an atomic nucleus—and it would likely be created moving faster than the escape velocity for the Earth. Such a black hole would sail right through the Earth out into space without absorbing anything more than a few protons or neutrons. But long before that, it would evaporate."

"Oh, right, Hawking radiation!"

EVERYTHING GOES AWAY IN THE END: EVAPORATING BLACK HOLES

Black holes are such exotic objects that they have captured the popular imagination, featuring in numerous novels and movies, as well as any number of popular science books. The way they push the limits of general relativity makes them equally fascinating to scientists, and black holes have been an active area of study more or less since Schwarzchild's initial work in 1915.

Research on black holes took a particularly dramatic turn in the 1970s. Scientists studying the equations governing the behavior of black holes noticed similarities between those equations and the equations of thermodynamics. Israeli physicist Jacob Bekenstein even proposed that the similarity was an exact correspondence, with a black hole having an entropy proportional to the surface area of its horizon. This was a rather bold assertion, and a number of people set out to prove him wrong, among them a young British physicist named Stephen Hawking. After a couple of years, Hawking shocked everyone by not only confirming Bekenstein's claim about the entropy but extending it to say that a black hole also has a temperature, which is inversely proportional to its mass: doubling the mass cuts the temperature in half.

To a nonphysicist, this may not seem shocking, but it was as radical a proposition as anything in the history of black hole physics. For the usual laws of thermodynamics to apply to black holes, black holes would have to behave just like any other object with a temperature. One of the most important features of thermodynamics is that hot objects emit radiation—the familiar red glow of a hot object being the best-known example. Explaining the origin of this radiation led to the development of quantum mechanics, so thermal radiation is an essential part of physics. And yet, emitting thermal radiation seems to go against the very idea of a black hole.

But that's exactly what Hawking showed: black holes *do* radiate, exactly like any other object. The temperature for a typical black hole is extremely low—0.00000003 degrees above absolute zero for a hole with twice the mass of the sun—but the equations show that it is real. More than that, Hawking showed that emitting this radiation causes a black hole to evaporate—little by little, it loses mass and eventually disappears in a burst of radiation.

How can this be? It's a consequence of quantum electrodynamics, which we talked about back in Chapter 8, specifically the virtual particles that constantly pop in and out of existence everywhere in the universe. Most of the time, they only have a tiny effect. But the event horizon of a black hole is no ordinary location in spacetime, and putting virtual particles into our model of a black hole leads very naturally to the radiation predicted by Hawking.

When a virtual particle–antiparticle pair pops into existence near the horizon of a black hole, one of the pair may be closer to the horizon than the other. If the tidal forces are strong enough, that particle can be drawn inside the horizon before it can annihilate with its partner. That partner, in turn, escapes to become a real particle, which an observer outside the black hole sees as radiation emerging from the vicinity of the horizon.

Working out all the details is difficult, but we can understand the qualitative behavior from the basic properties of virtual particles and black holes. The separation between a pair of virtual particles increases with their wavelength, so long-wavelength particles, which have low energy, are more likely to be separated by enough for tidal forces to pull them apart. That means that the radiation from a large black hole, with relatively weak tidal forces,* consists mostly of low-energy particles, as is characteristic of cold objects. A smaller black hole, with much sharper curvature and thus greater tidal forces, radiates more high-energy particles and thus has a higher temperature.

A black hole, then, is an object whose temperature increases as its energy content decreases. As the Hawking radiation whittles it away one particle at a time, the temperature of the black hole increases, and the radiation carries energy away more quickly. The evaporation of a smallish black hole with twice the mass of the sun would take a mind-bogglingly

---

* A large black hole has a large Schwarzchild radius, so its horizon is far from the singularity, where the tidal forces aren't that large.

long time—something like 10^{66} years—but the evaporation gets faster as it proceeds, with the last few thousand tons exploding out in a fraction of a second.

"I still don't understand this, dude. How does Hawking radiation make the hole evaporate? I mean, it eats a particle—shouldn't that increase the mass?"

"It eats a *virtual* particle, which is just a temporary manifestation of the vacuum energy. In a sense, the mass of that particle is energy that's temporarily borrowed from the vacuum. Ordinarily, it annihilates quickly, returning to vacuum energy. Having one particle become real and the other increase the mass of the black hole would increase the total energy content of the universe, which can't happen spontaneously. Something has to lose mass to keep the energy of the universe constant, so it comes out of the mass of the black hole."

"So, if nothing that falls into the black hole ever comes out, but a black hole emits Hawking radiation, does that mean that a black hole just turns stuff into random radiation?"

"Very, very slowly, yes—which is one of the interesting problems in black hole physics at the moment. This is the 'black hole information paradox': What happens to the information about the objects falling into the black hole? Are their states somehow contained in the Hawking radiation, and if so, how?"

"'Information' meaning, like, what particles are there and how they're put together?"

"Exactly. A lot of physicists lately think about physics largely in terms of information. So, a solid object is not just a collection of particles but a collection of information."

"But once something falls into a black hole, we lose the information about how it's put together, because we can't see inside the horizon, right?"

"Exactly. If the black holes lasted forever, the information would just be hidden inside, but they evaporate. And what comes out is Hawking radiation, which comes from random vacuum fluctuations and doesn't seem like it carries any information. That's a problem, because if information is fundamental to physics—and it certainly seems to be in all other areas—

then the total amount of information in the universe should not change, so an evaporating black hole *can't* destroy information. So, either information isn't fundamental, or it must somehow be carried in the random particles of the Hawking radiation."

"That's a puzzle all right. What's the solution?"

"It's not really settled yet. Hawking had bet some other scientists that information *is* destroyed in black holes, but he recently changed his mind and conceded the bet. Not everyone believes the argument he says changed his mind, though, so it's still unsettled."

"I'd like to suggest a third possibility, dude."

"Which is?"

"Black holes don't really exist. They're just a mathematical curiosity, not something in reality. This has two advantages: it gets rid of the information problem, and it also means I don't have to worry about black holes, which make my poor, fuzzy head hurt."

"That used to be a popular solution, but it's not anymore. These days, physicists and astronomers are pretty positive they exist."

"Yeah? Why?"

"Well . . ."

IT'S BETTER TO BURN OUT THAN TO FADE AWAY: STELLAR COLLAPSE AND BLACK HOLES

The idea of black holes is sufficiently weird and troubling that no sooner were they predicted than physicists started trying to rule them out. Any number of eminent physicists and astronomers tried to show that black holes could not possibly form in reality, but all their arguments fell short. Physicists and astronomers now accept black holes as an inevitable product of the death of very large stars.

A star, as the song has it,* is a mass of incandescent gas, mostly hydrogen, held together by its own gravity. Gravity attempts to pull all of the

* "Why Does the Sun Shine (The Sun Is a Mass of Incandescent Gas)," a 1959 song covered by They Might Be Giants in 1993 and 2009.

gas in a star down to a single point, but as the gas is compressed, its temperature rises, eventually becoming hot enough for hydrogen to fuse into helium. The fusion reaction releases energy, as we discussed in Chapter 7, keeping the core of the star hot and providing outward pressure that holds off the gravitational collapse.

As a very large star ages, the hydrogen fuel in the core begins to run out, and the star starts fusing helium, producing carbon and oxygen. The energy released in this reaction is slightly less, so the core contracts. When helium runs low, carbon and oxygen fuse into heavier elements, and so on, until the reactions produce iron. Iron can't release energy through nuclear fusion, so when the star gets to this point, the fusion reactions stop. Without the energy produced by fusion, there's nothing to withstand the effect of gravity, and the core implodes. The outer layers of the star are blown off in a titanic explosion, called a supernova, which can temporarily outshine the rest of the star's home galaxy.

What happens next depends on the mass of the core. If the imploding core is only slightly heavier than our sun, the collapse can be halted by nuclear interactions. As the pressure due to gravity increases, protons and electrons in the core of the star are squeezed so tightly that they combine to form neutrons.* The neutrons resist being packed tighter, which can halt the collapse, if the gravitational force isn't too strong. If the star is above 1.4 times the mass of the sun,** but not more than about 3 times the mass of the sun,† the collapse stops at this point, leaving a **neutron star**.

---

* This process involves turning up quarks into down quarks and releases a vast number of neutrinos, which should account for more than 90 percent of the energy released in the supernova. The neutrino pulse from a nearby supernova in 1987 was detected by experiments here on Earth and helped verify this model.

** This is known as the Chandrasekhar limit after the Indian physicist Subrahmanyan Chandrasekhar, who worked it out while traveling from India to England in 1930. Chandrasekhar was so distressed by a clash with his mentor Arthur Eddington over black holes that he stopped working on them altogether for decades. He only returned to black hole physics in the 1970s but still wrote the definitive textbook on the subject.

† The true upper limit on the mass of a neutron star isn't known precisely, but by three times the mass of the sun, it's definitely impossible to stop the collapse into a black hole.

If the mass of the collapsing core is greater than three times the mass of the sun, not even neutron pressure can withstand the force of gravity, and the collapse continues unabated, squeezing all the mass of the core down to a single point, which becomes the singularity at the core of the black hole. As the size of the collapsing core passes the Schwarzchild radius, the core is forever cut off from the outside universe by the event horizon. Nothing known to modern physics can prevent the formation of a black hole for a collapsing core heavier than three times the mass of our sun.

"You know, dude, this is all still theory. I mean, it's great that your models don't provide for anything that would stop a black hole from forming, but your models are just models. Has anybody actually observed these things?"

"I was just getting to that, before you interrupted. I warn you, though, the evidence is all going to be somewhat indirect."

"How's that?"

"Well, because, by their nature, neutron stars and black holes don't produce much light. So while we're fairly confident that we have detected them, we mostly see them through their effects on other objects."

"Didn't we just talk about how black holes radiate?"

"Yes, but the Hawking radiation from a typical black hole is way too faint to detect unless you're right on top of it. And since it's impossible to travel the vast distances to the nearest black holes, we can only deduce their existence from other evidence."

"Well, OK. But I still say you ought to go see one up close."

"Believe me, it's on the list if anybody finds a magic way to get there."*

How do we know that these stellar remnants really exist and aren't just figments of physicists' mathematical imagination? In the case of neutron stars, we directly observe them as **pulsars**, sources of short pulses of radio waves that repeat in a very regular way, which were first detected in the

* Don't hold your breath waiting for a magic travel method to emerge, however, because, as we discussed in Chapter 6, everything we know about relativity tells us that faster-than-light travel is impossible.

late 1960s by Jocelyn Bell Burnell. These pulses are the result of the gigantic magnetic field associated with a rapidly rotating neutron star. As the star spins, the magnetic field accelerates nearby electrons, producing powerful beams of radiation from the magnetic poles of the star. The magnetic poles of the star, like the magnetic poles of the Earth, are not necessarily located on the axis of rotation, so the pole sweeps past us once per rotation, like the rotating light atop a police car or lighthouse. We detect one pulse per rotation of the star—the slowest take several seconds to complete one rotation, the fastest mere milliseconds—and measurements of this rate and how it changes allow astronomers to determine how the neutron stars powering pulsars behave.

Black holes are more difficult to detect, because even a rotating black hole does not produce a magnetic field, thanks to its extreme gravity, and, of course, no light emerges from inside the event horizon. We can still detect black holes, though, using radiation produced just outside the horizon as matter falls into it. Gas typically does not fall straight into a black hole but follows a spiral path, moving faster and faster as it falls, like water rushing down a drain. As the gas speeds up, collisions between molecules heat it to the point where, just before it vanishes forever, it emits X-rays. Somewhat counterintuitively, then, we can detect black holes, which themselves emit no light, as bright sources of X-rays. Dozens of such sources have been detected since the 1960s, when X-rays from space were first observed from high-altitude rockets.

"Wait a minute, dude. All that tells you is that there are things that emit lots of X-rays. That could be lots of things that aren't black holes."

"Such as?"

"Ummmm . . . they have X-ray machines at the vet. Maybe there are lots of aliens with sick pets?"

"Nice try. But you're right, there are things other than black holes that could produce X-rays, which is why you need some additional information, like the mass of the object. If you see an X-ray source that is very small and has a mass greater than about four times the mass of the sun, you can be fairly confident that it's a black hole."

"How do you get the mass, though?"

"You can't easily get the mass of a single object, but conveniently for astronomers, many stars form as binary star systems, or two stars orbiting one another. When one of those turns into a black hole, the other continues to orbit it, and the rate at which it goes around depends on the mass of its partner. So, if you see a bright X-ray source being orbited by an ordinary star, you can determine the mass of the X-ray source by how fast the star is moving. Dozens of such systems are known in the Milky Way, most with masses between ten and twenty times the mass of the sun."

"And that's the size of a black hole?"

"An ordinary black hole, yes. There are supermassive black holes at the center of most galaxies, too, with masses of millions or even billions of times the mass of the sun."

"And you know that, how?"

"Well . . ."

The X-rays produced by an ordinary black hole consuming gas are pretty impressive, but they are by no means the most impressive effect attributed to black holes. Many distant galaxies are what astronomers blandly term "active galaxies," producing huge amounts of light from small regions at their centers. Many of these galaxies also have "jets," vast clouds of gas shooting out from the center of the galaxy at nearly the speed of light. The energy involved in producing these jets can be greater than the energy output of all the ordinary stars in the galaxy combined.

The only viable source for these enormous jets is a supermassive black hole, with a mass millions of times the mass of the sun. As such a black hole gobbles up gas in a young galaxy, it generates vast amounts of radiation and gigantic magnetic fields, which in turn accelerate some of the gas, producing the jets seen in most active galaxies. As time goes by, these black holes eventually eat up most of the free gas in their immediate neighborhood and stop producing so much radiation, but astronomers believe that most galaxies harbor supermassive black holes at their center.

"Those sound scary. I'm glad our galaxy doesn't have one."

"Actually, we do. There's a supermassive black hole at the center of the Milky Way."

"What? There is? Why doesn't anybody tell me these things?!"

"You didn't ask until just now. Anyway, it's nothing to worry about—it's a hundred thousand or so light-years away and pretty quiet as such things go."

"How do you know it's there, then?"

"Well, it does produce some radiation that we can detect. Mostly, though, we know because we can observe the orbits of stars going around the center and determine the mass of the thing they're orbiting."

"This is a Doppler effect measurement?"

"Actually, no. The stars close to the galactic center are moving so fast—up to 12,000 km per second—that they have moved substantially during the last decade or so of observations. Astronomers can directly trace out their orbits and show that they're orbiting something with 2.6 million times the mass of the sun."

"Yeah, but couldn't it be—"

"I know what you're going to say, and even as dense as the core of the galaxy is, there isn't enough room for 2.5 million extra stars. The only way to fit that much mass in that little space is a supermassive black hole."

"Oh, okay, I guess."

The idea of a black hole that warps spacetime so severely that not even light can escape is so strange and troubling that it has stimulated huge amounts of research over the ninety-odd years since the phenomenon was first predicted. As black holes represent the most extreme end of general relativity, this research has also proven extremely fruitful in developing new mathematical techniques for solving and interpreting the equations of general relativity.

Those decades of research have shown that, contrary to the hopes of many of the physicists who first began studying them, black holes do exist and are the inevitable end product of certain types of stars. Astronomical observations over the last few decades have confirmed these predictions and shown that as disturbing as they may be, black holes are fairly common and even essential objects in the universe.

"I don't know, dude. All you've really shown is that there are really, really heavy things out there. You haven't demonstrated any of the really cool space-warping stuff."

"True, but that's because you can only really see those effects by probing spacetime immediately around the black hole. Since none are close enough to visit, we have to make do with indirect measurements, which mostly involve things outside the black hole where the warping of spacetime isn't too severe."

"Don't you have *any* proof of the space-warping stuff?"

"You mean, other than the dozens of experiments in the solar system that confirm general relativity in great detail?"

"Yeah, other than that."

"Well, at least you're not making any unreasonable demands . . ."

"Just answer the question, dude."

"We can't directly measure the curvature of spacetime in distant places, but some observations of pulsars show the influence of **gravitational waves**."

"Waves? Like the water waves in the pond out back?"

"Not exactly, but the basic idea is similar. General relativity says that spacetime is flexible and can support disturbances that propagate like waves—they ripple outward from the source, stretching and compressing spacetime as they go."

"OK, that's just freaky."

"Astronomers have found a couple of really extreme systems—two pulsars orbiting each other very rapidly—that show the influence of gravitational waves. As they orbit, the pulsars are slowing down, which tells us that some of their energy is being carried away in a form that we don't detect directly. The only thing that could be is gravitational waves."

"So, you've detected them by not detecting them? That's kind of weak, dude."

"Maybe, but it ought to be possible to detect them directly, and in fact large experiments are under way now to look for gravitational waves. The biggest is the Laser Interferometer Gravity-Wave Observatory, or LIGO. They're hoping to measure the stretching and compressing caused by a passing gravitational wave using what is basically the world's largest Michelson interferometer."

"How big are we talking, here?"

"The arms are 2.5 km long, but they use mirrors to make the effective distance even longer. They can measure a change in that distance of about a tenth of the width of an atomic nucleus."

"This sounds pretty ambitious. They actually built this?"

"Two of them, one in Washington State and the other in Louisiana. That way, they can rule out random noise that affects just one of the detectors—if both see the same signal at the appropriate time,* they know it's real and not just a minor earthquake or something shaking the mirrors."

"And what have they seen?"

"Nothing, but then LIGO hasn't reached its full sensitivity yet. In the next few years, though, it should start detecting gravitational waves, opening a whole new way of looking at the universe."

"Including black holes?"

"Almost certainly."

"Well, I guess that's something to look forward to then. So, what else have you got?"

"What do you mean?"

"I mean, what other cool things can you do? Black holes are neat and all, but what else do you have to offer?"

"Well, black holes are about as extreme as it gets in this universe. There's only one bigger application, which is the universe itself."

"You can use general relativity to describe the entire universe?"

"Its origin and eventual fate, yes—in the next chapter."

---

* Gravitational waves should travel at the speed of light, so depending on the direction of the source, there should be a small difference in the arrival time in Washington versus Louisiana.

Chapter 11

EVERYTHING RUNS AWAY: GENERAL RELATIVITY AND THE EXPANDING UNIVERSE

I'M SITTING IN THE RECLINER READING a book about astronomy when Emmy comes up and nudges my elbow with her nose. "Dude, we need to talk," she says.

"What's the matter, girl?" I say, closing the book. "You look concerned." I scratch her ears.

"I'm concerned about the expansion of space."

"Ah." I put the book down. "That is a big problem and one that's bothered lots of scientists."

"Really?"

"Absolutely. The idea of spacetime itself changing is one of the strangest in general relativity. It takes a lot of getting used to."

"That is pretty weird, but—"

"Einstein himself had trouble with the whole business, which led to what he termed his 'greatest blunder.'"

"Wait, he made mistakes?"

"Lots of them. He spent years getting general relativity wrong before he got the final theory put together. And the last twenty-odd years of his life were spent in a futile argument against quantum theory. It's just that his mistakes were often brilliant, where ours are just dumb."

"Like forgetting to screw on the top of the human puppy's bottle before you shake it up?" Licking the spilled milk up was the highlight of her whole morning, and I expect I'll be hearing about that one for a while.

"Actually, I'm fairly certain that Einstein did that sort of thing. I'm talking about scientific mistakes, though, not general absentmindedness."

"Oh. So what is a brilliant scientific mistake like?"

"Well, according to general relativity, spacetime is sort of like an elastic material and can expand and contract depending on what happens with the matter it contains."

"The rubber sheet thing again."

"Exactly. The presence of matter bends spacetime, so if you try to measure the distance in space between two objects near the sun—Earth and the planet Mercury, say—you find that the distance is a little longer when the path you're measuring passes near the sun than when it doesn't pass quite so close."

"Because space stretches near the sun?"

"Time as well, but that's the idea. You're getting the hang of this stuff."

"That's because I'm the best!" She flops over on her back, and I rub her belly. She thumps her tail happily.

"Anyway, the point is that spacetime itself can expand and contract. Einstein realized this in 1917, when he started to apply general relativity to the universe as a whole, which was the birth of modern **cosmology**. He realized that his equations suggest that spacetime should either be expanding, if there isn't enough matter in the universe to stop it, or contracting, if the amount of matter is big enough to pull the whole thing back together. He found this awfully disturbing."

"I bet. So, that was his blunder?"

"No, the thing he later called a blunder was how he decided to fix it. He realized that if he stuck an extra term into the equations—which is perfectly allowed by the mathematical rules—he could arrange it so that the size of the universe was fixed and unchanging. Since most people in 1917 thought that was the case, this made the theory much more satisfying to him. The extra bit is known as the **cosmological constant,** because it has the same value everywhere in the universe, and it just balances things out, making spacetime static and unchanging."

"But now we know better?"

"Exactly. Not that long after Einstein's introduction of the cosmological constant, Edwin Hubble showed that the universe *is* expanding. Distant galaxies are rushing away from us, and the farther away they are, the faster they're moving. That's exactly what you'd expect to see if spacetime as a whole is expanding. Hubble's observations were the first evidence for modern **Big Bang cosmology,** which has the entire observable universe starting from an infinitely dense point at a particular instant in the past."

"So there wasn't any need for the cosmological constant? Einstein just added a fudge factor to his equations because he felt like it?" She looks smug. "I guess that *is* a blunder."

"It may seem funny, but nature always gets the last laugh. The cosmological constant turned out not to be necessary for the purpose Einstein intended, but it's made a comeback. You see, astronomical observations in the last fifteen years have shown that the expansion of the universe is *accelerating.* That is, spacetime is expanding faster than you would expect if the expansion were just a relic of the Big Bang."

"What's that have to do with Einstein?"

"Well, he stuck in the constant because it counteracted the tendency of the universe to collapse due to the matter in it. But that same constant can also, if its value is a little bigger, make the expansion accelerate—which is exactly what we see."

"So Einstein's blunder wasn't really a blunder?"

"Or, if you like, his real blunder was in thinking that inserting the cosmological constant was a blunder."

"That's awfully meta."

I shrug. "Whatever you like."

"So, what causes this constant, anyway?"

"An excellent question." She wags her tail proudly. "I wish I had an excellent answer for you, but the fact is, we don't know. The constant describes a sort of energy associated with empty space, which means that about 70 percent of the energy content of the universe comes from this **dark energy.** And we don't have a good explanation of why it has the value it does."

"Isn't that kind of a problem? I mean, we're all being stretched out by this mysterious force, and you don't have any idea what it is, let alone how to stop it?" She looks really distressed. "What kind of twisted, Kafkaesque universe is this, anyway?"

"OK, first of all, *we're* not getting any bigger. Atoms and molecules are bound together by very strong forces, so they stay the same size. It's only when we look across vast distances of empty space that we see the expansion at all."

"Well, that's a relief."

"And second, the expansion is very slow. It won't produce any visible effect until billions or trillions of years from now. So you and I don't need to worry about it at all."

"OK. You can't stop it, though?"

"Since it permeates the entire universe, I think the odds of human scientists eliminating dark energy are pretty minimal."

"Well, do me a favor and think about it, will you?"

"Sure. I'll let you know if I come up with anything." I rub her belly some more. "So, does that answer your questions about the expansion of spacetime?"

"Well, the cosmology stuff was interesting, but it hasn't really addressed my concern about the expansion of space."

"You mean spacetime."

"No, I mean *space*. As in, the space occupied by your puppy's toys. They're expanding to crowd me out of my own territory, and I don't like it. Not one bit."

I look around the room, and she's right. There are toddler toys everywhere. I sigh. "OK, I guess I can pick up a little."

"Thanks. And as long as you're getting up, how about you let me outside, hmm?"

To see truly dramatic effects of general relativity, we need to look at ever more massive objects, which is why discussions of relativity always involve black holes. If those aren't extreme enough, though, there's nothing more massive than the entire universe. Physicists and mathematicians wasted little time before applying general relativity to the universe as a whole, which started a process of discovery that continues to surprise us today.

In this chapter, we talk about cosmology, the branch of science dedicated to explaining the origin and history of, well, everything. We talk about the history of attempts to apply general relativity to the universe as a whole and about what theories and observations tell us about the origin and eventual fate of everything around us. And we talk about the recent observations telling us that for every bit of the universe whose nature we do understand, there are twenty-four bits that we don't.

EVERYTHING EVERYWHERE IS CURVED: SPACETIME AND THE EXPANDING UNIVERSE

In 1917 Einstein, never one to think small, turned to the question of applying his new theory of gravity to the universe as a whole. This might seem like an absurd proposition, since we don't know the exact position and mass of every object in the universe. As long as you're only concerned with the big picture, though, the problem is actually easier than working out the exact behavior of a collection of objects with known masses and positions. Finding the shape of spacetime on the scale of the entire universe becomes tractable if you make one simplifying assumption: that matter is distributed through the universe in an approximately uniform way. You don't need it to be perfectly uniform—matter can still clump together to form stars and galaxies and so on—you just need the probability of finding some amount of matter in any randomly chosen volume of space to be about the same everywhere.*

Once you make this assumption, which astronomical observations show is a reasonable approximation for our universe, it's not that difficult to apply the equations of general relativity to a model universe, which is what Einstein did in 1917. He immediately ran into a problem, though: according to the theory, the universe should be either expanding or collapsing. All of the matter in the universe bends spacetime in a way that attracts other matter, and given enough time, this should cause everything to collapse inward to a single point. You can avoid this problem by sending all the mass flying outward at high enough speed that gravity can't bring it back, but that requires everything to start at a single point, which isn't any more satisfying—avoiding a collapse requires a very particular distribution of velocities, and the end result is an empty universe, with individual clumps of matter all alone, infinitely far away from everything else.

The same problem occurs with Newton's theory of universal gravitation, which also has every bit of matter in the universe attracting everything else. When questioned about this very issue, Newton fell back on divine intervention, asserting that God would make sure the universe kept running smoothly.** This was just barely acceptable in Newton's day, but by the early twentieth century, the practice of attributing unknown causes to the agency of God had fallen out of scientific favor. Einstein himself had beliefs that are often characterized as religious, but they did not allow for the kind of activist God needed to keep the universe from collapsing or exploding.

---

* Much of Einstein's physics education and early scientific work was in the area of statistical physics, where this sort of approximation is very common. As a result, he was already familiar with the tools used to handle this kind of situation mathematically, making this a natural problem for him to attack.

** Newton was intensely, if somewhat eccentrically, religious and attributed many things to God.

Mathematics, however, provided a way out of the problem. Einstein realized that he could fix the problem of universal collapse by adding an extra term to his equations. This cosmological constant, commonly represented by a capital Greek lambda (Λ) describes a sort of intrinsic energy of empty space, causing it to naturally expand outwards. It adds a small amount of energy to every point in the universe, but because the extra energy is the same everywhere, it doesn't affect the observable behavior of ordinary objects. By setting this constant to the right value, Einstein could exactly balance the inward pull of gravity, resulting in a uniform and unchanging universe. As this seemed both philosophically satisfying and consistent with astronomical observations, he published his relativistic theory of the universe including the constant.

"Wait a minute. How is that any better than invoking God?"

"What do you mean?"

"Well, the size of this constant would need to be just right to match the total mass of the universe, right?"

"Yes, that's right. The static universe only works if the constant has a pretty specific value."

"So, how is that any better than having God pushing things around? I mean, this extra mysterious constant just happens to have the right value to make everything look nice? How does that make any sense?"

"It's not an ideal solution by any stretch—Einstein himself was never all that happy with the idea. But it's really not that much different from the situation we always have in physics. There are tons of specific values that we use all the time that seem arbitrarily chosen. I mean, why is the mass of an electron what it is? Or the charge on a quark? Or the strength of gravity? All of those are set by arbitrary constants whose origin we don't know but just have to accept. From that perspective, adding one more arbitrary value that just happens to make things work nicely doesn't seem that awful."

"Yeah, but there's no reason for it. I mean, we can measure gravity and mass and the rest. But just sticking in a fudge factor to make things work out seems so . . . bush-league."

"True. And it wasn't long before people started asking what would happen if you left that factor out."

"As they should."

Einstein's model of a stable and unchanging universe agreed with what astronomers knew in 1917, but it's not the only possible model. It didn't take long for other scientists to explore the other possibilities. In 1922, Russian cosmologist Alexander Friedmann used Einstein's equations without the cosmological constant to show that they provided three possible histories for the universe, each depending on a particular shape of spacetime.

If the amount of matter in the universe is sufficiently large, the universe starts out very small and expands to a huge size, but the expansion slows down because of the gravitational attraction of all that matter and eventually reverses, with the entire universe collapsing back to a single point. If the amount of matter in the universe isn't big enough, the universe starts small, then expands to huge size, keeps expanding forever, becoming infinitely large. In between those two possibilities, there's a very particular amount of matter in the universe, the **critical density,** the Goldilocks "just right" situation: the universe starts small and expands, but gravity slows that expansion down. The expansion never completely stops but keeps getting slower and slower as time goes on. Friedmann showed that these three cases cover all the possible solutions of Einstein's equations—the expansion may be a little faster or a little slower in places, but these three cases cover all the possible fates of the universe.

The common feature of all three of these models of the universe is that its size is always changing in time. In the early 1920s, there was no solid reason to expect this kind of dynamic universe, so many scientists rejected Friedmann's findings. Einstein particularly hated them and even tried, unsuccessfully, to find a mathematical flaw, before conceding without much grace that they were mathematically sound. He continued to reject the underlying idea, though, holding to the idea of a static and unchanging universe.

"So, what does this have to do with bending space? How did this Friedmann guy get these silly results in the first place?"

"Well, general relativity tells you that the presence of matter bends spacetime, so the matter in the universe bends it into a particular overall shape. Mathematically, these shapes can be categorized by a single number, the curvature of spacetime, and there are only three types of values that curvature can take: positive, negative, or zero."

"That would seem to cover all the bases, yes."

"A universe with a lot of matter in it has positive curvature, which is the sort of curvature you find in the two-dimensional surface of a sphere. In a space with positive curvature, you find that parallel lines are drawn together and eventually cross."

"Hang on—parallel lines don't cross. That's what makes them parallel."

"Parallel lines don't cross *in flat space*, but on a curved surface they do. Think of two lines of longitude on the surface of the Earth: near the equator, they look absolutely parallel, but if you follow them along far enough, you find that they eventually come together at the pole."

"So if you have enough matter, space bends so that when you head off in one direction, you eventually come back to where you started? How does this mean that the universe started small and will get small again?"

"Spacetime bends. Remember, the universe has four dimensions, one of which is time. The complete shape of the universe is a four-dimensional object that isn't easy to picture but has the same mathematical properties as the surface of a sphere. If you want to picture the universe as the surface of a sphere, you should imagine that one of the two dimensions of the surface is time. So, for example, you can think of lines of longitude as marking out space, while lines of latitude mark the passage of time."

"So, the universe begins as a really small thing at the South Pole, becomes big as you move up in latitude, then becomes small again and ends at the North Pole?"

"An Australian might flip the order of the poles, but yes, that's the idea. The lines of longitude represent the paths through spacetime of objects that aren't moving, so they start out flying apart, then come back together."

"OK, so that's positive curvature. What are the other two?"

"Negative curvature describes the model where the universe starts small and just keeps expanding. It's harder to picture, but the standard example of a surface with negative curvature is a saddle. If you draw two lines that are parallel up on the seat of the saddle, as you move out to one side or the other, those lines move farther apart. So, in the analogy to the history of the universe, two things that are some distance apart now end up much farther apart in the future."

"Yeah, but a saddle doesn't start at a particular point, so how is it expanding?"

"Like I said, it's an analogy. The real shape of a negative-curvature spacetime is a four-dimensional object, not two-dimensional like the surface of a saddle, and if you work it all out, you find that it starts small and keeps expanding forever."

"How do you picture a four-dimensional surface, anyway?"

"Oh, that's easy. You just imagine it in *N* dimensions, then let *N* go to four."

". . ."

"That's an old math joke. I've never been able to get the hang of it, myself, which is a big part of why I'm an experimental atomic physicist and not a general relativity theorist."

"You could just *say* that, you know. You don't need to mess with my head all the time."

"Sorry."

"All right, that's two of Friedmann's universes. What's the third?"

"The final possibility is a zero-curvature, or flat, universe. In a universe with zero curvature, the usual rules of flat-space geometry hold: parallel lines don't cross, the angles in a triangle add up to 180 degrees, etc."

"See, now this option I like."

"It does have its appeal. Friedmann showed that you could have a universe in which spacetime as a whole was flat, but the universe evolved in time, starting out small, expanding at a decreasing rate but never turning around."

"That's weird, but I guess it makes sense in four dimensions?"

"Pretty much."

"So, how did Friedmann get this flat universe?"

"Well, the Friedmann model of a flat universe requires the total amount of matter in the universe to have exactly the right value."

"Isn't that an awfully big coincidence?"

"Yes, but remember, Einstein's static model required the cosmological constant to have exactly the right value to counter the effect of the matter in the universe. So they both have arbitrary and huge coincidences."

"Hmmm . . . I guess so. So, which of them was right?"

"Well, neither. Or both, depending on how you look at it."

Friedmann continued to press his model of a dynamic universe in spite of the objections of Einstein and other prominent scientists, but sadly, he contracted typhoid fever and died suddenly in 1925. The notion of an expanding universe was independently rediscovered by Belgian astronomer Georges Lemaître, who took the idea even further than Friedmann, noting that expanding spacetime implied a moment of creation, an instant when the entire universe was contained in a single extraordinarily dense point. Lemaître, who, in addition to being a credentialed astronomer and a decorated World War I veteran, was a Catholic priest, found this idea of a "primeval atom" particularly attractive and worked through its implications with great enthusiasm. When Lemaître met Einstein at the 1927 Solvay Conference, he jumped at the chance to explain his theory. Einstein pointed out that Friedmann had already found the same results and then added, "Your calculations are correct, but your physics is abominable."

There is no small irony, then, in the fact that just six years later at a 1933 seminar in California, Einstein described a presentation by Lemaître as "the most beautiful and satisfactory explanation of creation to which I have ever listened." So, what happened between 1927 and 1933 to produce such a dramatic turnaround on Einstein's part?

The most concise answer to that question is Edwin Hubble. In 1924, building on the work of Henrietta Swan Leavitt, Hubble showed that the Andromeda Galaxy is well outside the Milky Way—around 2.5 million light-years away—and thus a galaxy in its own right, not merely a fuzzy object within our own galaxy. At a stroke, Hubble vastly increased the size of the known universe—many astronomers had believed that the Milky Way comprised the entire thing—which alone would have guaranteed him a place among the greatest astronomers in history.

Hubble wasn't done, though. He next turned to spectroscopy, measurement of the characteristic colors of light emitted by stars, which astronomers use to determine the composition of distant objects. In the mid-1800s, physicists had found that each element of the periodic table emits light in a very particular set of frequencies, which can be used to "fingerprint" elements in different materials.* This was quickly applied to the sun, where the element helium was first identified thanks to a set of unexplained spectral lines. Further spectroscopic measurements identified the same elements in the light of distant stars, though on close inspection, the frequencies of these lines were often found to be shifted slightly from the values observed on Earth.

This shift is due to the Doppler effect, which we discussed briefly in Chapter 9 when talking about the gravitational redshift. The shift in the light detected from distant stars is thus a measure of their velocity relative to the Earth: light from stars moving toward us is blueshifted, while light from stars moving away from us is redshifted. The shift is tiny—a source of green light moving at 1,000 m/s would have its frequency shifted by just 0.0003 percent—but it is easily measured using spectroscopic techniques.

In the decade before Hubble's discovery of the size of the universe, astronomer Vesto Slipher had measured the spectra of many "spiral nebulae" and found most of them to be substantially redshifted, indicating that they are moving away from us at high speed. Having shown that these nebulae are other galaxies, Hubble turned to investigating their velocities through spectroscopy. His study not only confirmed Slipher's observations but added another intriguing detail: not only are galaxies moving away from

* Explaining this observation was a far more difficult task and eventually led to the development of quantum mechanics, which is described in *How to Teach Physics to Your Dog* (New York: Scribner, 2009).

us, but the speed at which they are moving away depends on the distance. The farther a galaxy is from us, the faster it is moving.

Hubble's result seems shocking, but it follows naturally from Lemaître's expanding universe: if space itself is expanding, you expect to see all galaxies moving away and more distant galaxies moving away more quickly. We can see how this works by considering a pack of dogs: imagine that the dogs start out sitting at rest, spaced 1m apart; then, spacetime in the dogs' neighborhood expands to twice the size in just one second.

As Figure 11.1 shows, a dog in the center of the pack, watching the expansion, will see all the other dogs rushing away from her at a speed that increases with distance. Bodie and Anson, immediately to Emmy's left and right, were initially 1m away and are now 2m away, so they are moving away at 1 m/s. The next dogs out, Winthrop and Maeve, started out 2m away but are now 4m away, so they are moving away at 2 m/s. And so on—the farther out in the pack you go, the faster the dogs are moving away. This recession velocity is not due to the dogs, who are just sitting still, but is a consequence of the expansion of the universe itself.*

Figure 11.1.

* If you don't feel up to herding dogs and stretching spacetime, you can get the same effect by putting a series of dots on a rubber band and stretching it to twice its original size. From the point of view of any of the dots, all of the others appear to be moving away, just like the dogs in the example above.

"Yeah, but that's just because I'm at the center of the universe, so everything is expanding away from me. A dog on the periphery would see something different."

"You might think that, and it was one of the early objections to this model—that it's awfully unlikely that we would happen to be at the exact center of the universe—but that's not the case. The expansion produces the same effect for everyone."

"How can that be? I'm in the middle, so I see all the other dogs running away, but surely the other dogs see some of them headed in the same direction they are."

"That's what would happen if the dogs were running, but remember, they're each sitting still while spacetime expands around them. Look, we can redo the diagram from the point of view of one of the other dogs (see Figure 11.2). According to Winthrop two spots to your left, he's standing still, while you have moved from 2m to 4m away. Meanwhile, he sees Bodie moving away at 1 m/s, while Zoe, on his left, is moving left at 1 m/s. And Anson seems to be moving right at 3 m/s, and so on."

"So, expanding spacetime makes everybody think that they're the center of the universe, even though it's really me?"

Figure 11.2.

"The point is that there really isn't any unambiguous center of the universe. Every observer is equally entitled to make that claim, as they all see the expansion sending stuff flying away from them at a speed that increases with distance."

"So where's the real center? I mean, if everything exploded from one point, that point had to be somewhere, right?"

"Since spacetime itself is what's expanding, the question doesn't make sense. At the instant of the Big Bang, all places were the same place. So the center of the expanding universe is either nowhere or everywhere, depending on whether you're a glass-half-full or glass-half-empty type."

"I prefer to think that the glass is twice as big as it needs to be. I'm still not sure I buy this, though. If space is expanding, shouldn't all the dogs just get bigger, making it seem like the distance stayed the same?"

"That's another common point of confusion. While *space* is getting bigger, the *stuff* in that space doesn't expand. You and I are held together by the electromagnetic attraction between the atoms and molecules making us up, and that's more than strong enough to hold us together despite the expansion."

"So all those distant galaxies aren't really rushing away from us?"

"Not in the sense of moving *through* space, no. The distance between us increases as they move *with* space, but it's not a physical motion of the galaxy in the way that people usually mean."

"So, they're really just sitting there?"

"They do move through space as well, because gravity attracts them to each other. The motion through space is in addition to the Hubble expansion with space, and astronomers can keep track of that as well. For some nearby galaxies, like the Andromeda Galaxy, the motion through space is more important—Andromeda is actually headed toward us."

"Is it going to hit us and make a mess?"

"Maybe. It's probably moving sideways as well, though we can't measure that as easily. And keep in mind, galaxies are mostly empty space, so we're not talking a billiard-ball type of collision. Anyway, whatever happens won't happen for 9 to 10 billion years, so don't lose any sleep over it."

"Don't worry about that, dude. Nothing gets between me and my sleep."

IN THE BEGINNING WAS NOTHING, WHICH EXPLODED*: THE BIG BANG MODEL

Numerous scientists attempted to find alternative explanations for Hubble's results, but none of them held up. The best explanation for the expansion observed by Hubble is Lemaître's: distant galaxies are moving away from us because the universe used to be smaller than it is and has been expanding for the last several billion years. The case for expansion is so strong that just six years after calling it "abominable," Einstein completely reversed course, praising the expanding-universe models of Lemaître and Friedmann as "beautiful" and calling his own invention of the cosmological constant the greatest blunder of his life.

The expanding-universe model has been refined since the 1930s, but the core of the model, which now goes by the name Big Bang cosmology,** is not too far from Lemaître's primeval atom. All evidence points to a universe that started out as very small and unimaginably hot and has been expanding and cooling ever since. The cause of the Big Bang is probably unknowable, but we can date it to around 13.7 billion years before the present, when the current universe came into existence as an infinitesimal point and immediately began expanding. Modern cosmology also includes an "inflationary epoch" shortly after the Big Bang in which the universe expanded incredibly rapidly—increasing its size by a factor of at least 10^{26} (probably much more) in just 10^{-32} seconds.†

At the end of inflation, the universe was extremely hot and dense, and for the next several hundred thousand years, it expanded and gradually cooled. For most of that time, the entire universe was an opaque plasma

---

* A pithy description of the Big Bang from Terry Pratchett's novel *Lords and Ladies* (New York: HarperTorch, 1996).

** The name was given in scorn but adopted with pride: the term "Big Bang" started as a disdainful reference by the British astronomer Fred Hoyle, who maintained a belief in a "steady-state" universe into the 1990s. While Hoyle made important contributions to the understanding of stellar evolution and the creation of elements, his theories of cosmology were never widely accepted and now seem a little nutty.

† To put that in perspective, if you inflated the size of a single atom by a factor of 10^{26}, its nucleus would be the same diameter as the orbit of Venus.

of charged particles and photons. Light couldn't travel very far before being absorbed, while atoms were unable to form without high-energy photons blasting them back apart. As the universe expanded, though, it cooled off and thinned out, and around four hundred thousand years after the Big Bang, protons and electrons could finally combine into atoms without being immediately destroyed. At that point, the universe became transparent to light, and so it has remained ever since. Over the next 13 billion or so years, slightly denser patches of the universe attracted gas gravitationally, eventually coalescing into galaxies, stars, and planets. Eventually, on an insignificant rocky planet around an ordinary star in an unexceptional galaxy, humans and dogs evolved and began to understand the universe around them.

The best evidence we have regarding the Big Bang (other than Hubble's observations of galactic redshifts) comes from the moment when the universe first became transparent. The vast numbers of photons flying around the early universe at that time are still around today. As the universe has expanded, though, these photons have redshifted from the gamma rays and visible photons of billions of years ago and are now found in the microwave region of the electromagnetic spectrum.*

This **cosmic microwave background** was first predicted by Ralph Alpher and Robert Herman at the Johns Hopkins University Applied Physics Laboratory. Alpher and Herman realized that some of the original radiation from the Big Bang should still be around and estimated the wavelength at which it would be found, based on then current models of the age of the universe.** The predicted radiation could not be detected with 1949 technology, though, so they were unable to get astronomers to confirm their prediction.

---

* Microwaves have a wavelength of a few millimeters or more, compared to a few hundred nanometers for visible light and a small fraction of a nanometer for gamma rays.

** The work grew out of Alpher's PhD thesis research on the formation of elements during the Big Bang. The main result from his thesis is known to astrophysicists as the "αβγ" paper, because Alpher's advisor, George Gamow, added Hans Bethe to the author list so that their surnames would sound like the first three letters of the Greek alphabet. (Gamow was infamous for his quirky sense of humor.)

Microwave technology advanced over the next fifteen years, however, and around 1964, researchers at Princeton University began to construct a detector to look for the cosmic microwave background. Much to their chagrin, they got scooped by a pair of researchers at nearby Bell Labs, Arno Penzias and Robert Wilson, who had spent weeks looking for the source of a pervasive background noise in a giant microwave detector they were building for radio astronomy. After running through a huge range of possible mundane sources—everything from nearby electronic devices to the pigeons nesting inside their antenna—they realized that their annoying microwave noise was in fact the echo of the Big Bang.

Physicists characterize a broad radiation spectrum like the cosmic microwave background by the temperature of an ideal object that would emit such radiation. As all dogs know, hot objects emit light, and the higher the temperature, the shorter the wavelength. Objects like flames or lightbulbs have a temperature of a couple thousand degrees Celsius and emit light in the visible range of the spectrum (400–700 nm wavelengths), while objects at room temperature or a bit higher—around 300K (300 degrees Celsius above absolute zero)—emit light in the infrared region (1,000–10,000 nm wavelengths). The cosmic microwave background first detected by Penzias and Wilson has a temperature of 2.725K, not far from the 5K predicted by Alpher and Herman fifteen years earlier.

Prior to Penzias and Wilson's discovery, the Big Bang model was just one possible theory of the history of the universe. After Penzias and Wilson, the Big Bang quickly became established as the consensus model of the origin of the universe.* Observations of the microwave background have become more and more sophisticated over the last forty years, and measurements of its average temperature have become more sensitive. The radiation has also been shown not to be perfectly uniform: two NASA satellites, the Cosmic Background Explorer (COBE) in 1992 and the Wilkinson Microwave Anisotropy Probe (WMAP) in 2003,** showed that the tem-

* Penzias and Wilson shared the 1978 Nobel Prize for their discovery.

** There have also been a number of microwave background measurements made by smaller experiments carried aloft by weather balloons, but COBE and WMAP are the only experiments to cover the whole sky and thus attract most of the attention.

perature varies by about 0.00003 degrees from one part of the sky to another. These small temperature fluctuations reflect tiny differences in the density of the universe at the moment it became transparent. These density differences provided the seeds around which stars and galaxies formed, so press-release descriptions of the results tend to be either cutesy ("baby pictures of the universe") or grandiose ("like seeing the face of God"). However it is described, though, the cosmic microwave background is our best source of information about the early universe, and it is best explained by the Big Bang model.

"So, Einstein wasn't just wrong, he was spectacularly wrong?"

"Well, to be fair to him, before 1929, there wasn't evidence for an expanding universe, so adding the cosmological constant was perfectly legitimate. And to his credit, when better evidence appeared, he immediately and publicly changed his position."

"Still, he didn't exactly cover himself in glory, did he?"

"No, but he was only human."

"You hairless apes are so fallible."

"Says a dog who freaks out over dogs barking on TV."

"I'm just concerned that they allow such language on broadcast television. I mean, there could be puppies listening . . ."

"*Anyway*, as I said back at the beginning of the chapter, the funny thing about this whole episode is that Einstein was more right than he realized. The cosmological constant has made a comeback in recent years, in a big way."

"Oh, right. What's the deal with that again?"

"Well . . ."

A BRILLIANT MISTAKE: THE COSMOLOGICAL CONSTANT AND THE ACCELERATING UNIVERSE

Since Friedmann's work in the 1920s, physicists have known that they need only a few parameters to describe the universe as a whole. The most important of these are the curvature of the universe, discussed earlier, and the density of matter and energy, which is usually given as a fraction of the critical density for the universe to just avoid collapsing back together. The two quantities are related—if the total amount of matter and energy exactly matches the critical value, the universe will be flat (zero curvature)—but they can be measured separately. By measuring both of these properties, astronomers can say exactly what the universe contains and what its ultimate fate will be. These are the biggest of big questions; thus, they are of great interest to physicists and astronomers.

Many different methods have been employed to try to measure the curvature of the universe, the most successful of which looks at the size of the most distant (thus oldest) thing we see, namely, the cosmic microwave background. The COBE and WMAP satellites measured small fluctuations in the temperature of the microwave background that came from small differences in the density of the gas filling the universe when it first became transparent. Models of the early universe predict the range of sizes for these "hot" and "cold" spots, so the observed size tells us something about the curvature of the universe. If the spots appeared much larger than expected, that would suggest a positive overall curvature for the universe, pulling light rays together and making distant objects seem larger. If the spots appeared much smaller than expected, on the other hand, that would suggest a negative overall curvature. The hot and cold spots detected by WMAP fit the predicted size distribution almost perfectly, suggesting that the universe is very nearly flat. This fits well with other observations that have tried and failed to detect some overall curvature of the universe.

"Wait, I thought you said that light bends as it goes by the sun, and you've got gravitational lenses and stuff. So how can the universe be flat?"

"We're talking about the largest possible scales. Spacetime is curved locally—any object with mass bends it—but the universe as a whole is flat, as far as we can tell."

"Dude, how is that not a contradiction?"

"Well, think of the universe as being like our yard. Would you say our yard is flat?"

"Yeah, sure. It's not like we're on a big hill. It might slope a little from back to front, but if you set a ball down on the ground, it won't roll anywhere."

"Right—on a large scale, our yard is flat because the ground is more or less level. On a smaller scale, though, there are lots of variations. There's that little hole you've dug yourself in the shade of that one bush, and there's the pond, of course, which is lower than the rest of the yard. And on an even smaller scale, there are lots of places where tree roots have pushed the ground up a little and other spots where critters have dug little holes, and so on."

"So, you're saying that the universe as a whole is flat like our yard as a whole, but on a small scale, matter makes it lumpy, like the yard when I can't find a comfortable spot to lie in the sun?"

"Exactly. When cosmologists talk about curvature, they mean the curvature of the entire universe, not the local curvature caused by stars and planets and black holes."

"OK, I guess that makes sense. So, the universe contains exactly the right amount of matter to make space as a whole flat?"

"Not exactly . . ."

The flatness of the universe is related to the amount of matter and energy contained in the universe, so you might think that the observed flatness means that we live in a "Goldilocks" universe with just the right amount of matter. Unfortunately, the situation is much more complicated than that. When astronomers add up the mass of visible matter in the universe—stars, galaxies, clouds of gas—they get a number much smaller than the critical density. This problem has been known for decades, and many different attempts to account for the missing mass have come up short, with the largest estimates of the total being about 20 percent of the critical density.

In order for spacetime as a whole to be flat but the amount of visible matter to be small, the universe must contain a huge amount of matter and energy that we can't see. Given the vast array of ordinary objects that don't emit enough light to be seen with a telescope—cats, squeaky toys, peanut butter—this might not seem like that big a deal, but the problem is much worse than that. Theoretical models predict that the vast majority of the atoms in the universe should be hydrogen—which is what we see—with the exact proportion of hydrogen to heavier elements depending on the density of matter when atoms first formed. Astronomers can measure the amount of hydrogen compared to other things and determine the density of ordinary matter in the early universe; when they do, they also find a value well below the critical density, about the same size as the total from stars and gas. So, if there's enough matter and energy in the universe to make it flat, it can't be ordinary matter made up of quarks but has to be something more exotic.

Astronomers have long believed that there is a large amount of matter present in the universe that isn't visible to their telescopes. The idea was first floated as early as 1933 by Fritz Zwicky,* but didn't become generally accepted until observations by Vera Rubin in the early 1970s showed that stars in galaxies orbit faster than can be explained from the mass of the stars and other visible material. Combined with the measurements of light elements, this tells us that the vast majority of the matter in the universe is a mysterious substance called **dark matter** because it doesn't interact with light and astronomers aren't any better at naming things than physicists.

Our picture of the universe got even stranger in the late 1990s, though, when two different teams of astronomers, headed by Saul Perlmutter of Lawrence Berkeley National Lab and Brian Schmidt of the Mount Stromlo Observatory in Australia, determined the distances and redshifts of the most distant galaxies then observed, extending Hubble's measurements to dramatically greater distances. Measuring galaxies at greater distances also means looking farther back in time, because light from those galaxies has taken billions of years to get here. The most distant galaxies we see formed when the universe was only a few billion years old and the rate of expansion was much different. Conventional wisdom said the expansion was faster in the past, and the astronomers measuring these galaxies

---

* Zwicky argued that the observed redshifts of galaxies in clusters showed they were moving too fast to be held together gravitationally without some large amount of unseen mass. The observations were never quite good enough to be convincing on their own, though.

planned to use the slowing of the expansion to measure the matter content of the universe: the slowing would be caused by gravity pulling everything together, so knowing how much the expansion had slowed would tell us how much matter is present and attracting other matter gravitationally.

Shockingly, they found that rather than slowing down, the expansion of the universe is speeding up. Spacetime is expanding faster today than it was several billion years ago, as if some force is pushing everything outward. This result was completely unexpected but unmistakable: it was arrived at independently by two competing groups with no great love for one another* and has since been extended and corroborated by other observations.** The best modern measurements, based on the expansion and the WMAP microwave background, suggest that ordinary matter making up dogs and bunnies and physicists accounts for only 4 percent of the matter and energy in the universe. Dark matter, whatever it is, accounts for another 23 percent of the energy content, meaning that 73 percent of the energy content of the universe is an unexplained quantity termed "dark energy."

"Wait a minute. You mean scientists have absolutely no idea what 96 percent of the universe is made of?"

"'Absolutely no idea' is a little harsh—we do know some of the properties of the unknown stuff—but it's true that 96 percent of the energy in the universe is in forms that we do not understand in detail."

"Why should I trust anything you say about physics, if you're only right 4 percent of the time?"

"Because the experiments and observations showing the existence of dark matter and dark energy are really solid. It's not just one or two little things; it's a whole host of stuff. Dark matter shows up in the motion of stars orbiting the center of galaxies, in the way galaxies move around within clusters, in the way galaxies and clusters of galaxies bend light. It's the simplest explanation for a huge range of phenomena."

---

* Richard Panek's *The 4 Percent Universe* (Boston: Houghton Mifflin Harcourt, 2011) tells the story of the discovery and the scientific rivalry between the two teams.

** Perlmutter, Schmidt, and Adam Riess were awarded the 2011 Nobel Prize in Physics for this discovery.

"And yet, you don't know what it is."

"Well, yeah, we don't know what it is. There are a lot of theories for what dark matter might be, mostly involving as-yet-undetected types of particles. There are a bunch of experiments looking for those particles, but nobody has detected anything concrete yet."

"More giant particle accelerator stuff, right? Smashing things together and hoping to make new particles?"

"That's part of it, but there are lots of experiments trying to detect the extremely rare interactions between dark matter particles and ordinary matter. This doesn't happen very often, but given the vast amount of the stuff—for every 1 kg of ordinary matter in the universe, there are almost 6 kg of dark matter—it must happen occasionally, and there are ways to look for it."

"Such as?"

"Well, if a dark matter particle bumps into an ordinary atom, that atom picks up some energy, producing a tiny amount of heat. The Cryogenic Dark Matter Search experiment in the United Kingdom monitors a bunch of really cold silicon and germanium crystals, looking for a tiny increase in the temperature that would indicate a dark matter interaction."

"If you say that it's pretty cool, I might bark at you."

"There's also the XENON Dark Matter Search Experiment, which uses a large amount of liquid xenon to look for collisions where a dark matter particle knocks an atom apart. They look for the small flash of light and charged particles produced by such an interaction."

"So, have they seen anything?"

"Not yet, no, but they've only just started, and the interactions they're looking for are extremely rare. It'll be a while yet before they can say anything definite."

"OK, but what about dark energy? Do you have any idea what that is?"

"No, but again, there are some hints. It doesn't obviously affect motion on small scales but is pushing the universe outward on really huge scales. That suggests it's a property of empty space rather than collections of matter."

"Energy associated with empty space pushing things apart? That sounds sort of familiar . . ."

"It should, because we talked about it a few pages back."

"Oh, wait, the cosmological constant!"

"Exactly."

Dark energy is an extremely weird idea—some quantity associated with empty space that pushes things apart—but also a familiar one. We've encountered this idea before, with the cosmological constant: Einstein added it to his equations to provide exactly this sort of outward push, resisting the tendency of the matter in the universe to draw everything together. In Einstein's case, the constant was carefully set to balance gravity and keep the universe from changing. In the modern version, the value of the constant is larger than Einstein's ideal value and dark energy pushes the universe outward faster than gravity pulls it back. This gives the same effect as a negative-curvature universe with too little matter, but the total energy, including the dark energy of empty space, ends up being right for the universe to be flat.

The exact source of the dark energy isn't known any better than in Einstein's day. Quantum mechanics does predict that empty space contains some energy, as we talked about in Chapter 8, but the simplest estimates of that energy are bigger than the observed value of dark energy by an absolutely mind-boggling amount: something like 10^{120}. That's a trillion trillion trillion trillion trillion trillion trillion trillion trillion trillion, vastly larger than the total number of protons in the visible universe. The vacuum energy is almost certainly the source of the dark energy, but we don't yet understand exactly how to connect the two.

Unlike Einstein's introduction of the cosmological constant, though, its modern use is well justified. Dark energy in the form of a cosmological constant is far and away the best and simplest explanation for the flatness and accelerating expansion of the universe. As Sherlock Holmes famously said, "When you have eliminated the impossible, whatever remains, however improbable, must be the truth." As improbable as dark matter and dark energy seem, astronomical observations make all the alternatives impossible, forcing physicists to the uncomfortable conclusion that we know far less about the contents of the universe than we would like to.

"Wow. Sucks for you guys, I guess."

"Yes and no. It does mean that there's a lot yet to be done. But, on the bright side, it means there's a lot of cool stuff yet to be discovered. There's a saying often attributed to Isaac Asimov that fits fairly well: 'The most exciting phrase to hear in science, the one that heralds new discoveries, is not "Eureka!" but "That's funny . . ."' A weird and unexplained phenomenon like dark energy affects scientists the way bacon affects dogs: they flock to it and get all excited."

"Bacon?! Where? Baconbaconbacon!"

"It's just an analogy. There's no bacon."

"Darn. So, do you have any other possibly apocryphal quotes you'd like to share?"

"How about Newton: 'I do not know what I may appear to the world, but to myself I seem to have been only like a boy playing on the sea-shore, and diverting myself in now and then finding a smoother pebble or a prettier shell than ordinary, whilst the great ocean of truth lay all undiscovered before me.' With, you know, the 96 percent of the universe we don't understand playing the role of the ocean of truth."

"OK, I admit, that's pretty good."

Despite its fearsome reputation, the central idea of relativity is astonishingly simple: the laws of physics need to work the same way for moving observers as stationary ones. When applied to observers moving at constant speed relative to one another, this leads to the theory of special relativity, which predicts time dilation, length contraction, and the equivalence of mass and energy. When applied to observers who are accelerating or experiencing gravitational forces, it leads to the theory of general relativity, which predicts the stretching of space and time due to large masses. As we've seen through the course of this book, we can understand the implications of these ideas by considering surprisingly simple systems such as dogs and cats that are watching moving clocks or falling in elevators.

The resulting theory has a truly astonishing reach, though, from practical technologies like nuclear energy or the Global Positioning System to exotic objects like subatomic particles and black holes. And, as we have seen in this chapter, general relativity provides the essential framework for understanding the origin, history, and eventual fate of the entire universe. It's one of the greatest achievements of human intellectual history, and, amazingly, it all starts with the daydreams of a patent clerk.

"I like the Einstein reference, dude, but this is still a pretty depressing way to end the book."

"What do you mean?"

"Well, you've got this whole history of the universe that ends with us all being ripped apart by some dark thingy that nobody knows what it is. I mean, that's a huge downer."

"First of all, we're not going to be ripped apart by dark energy any time soon. The most common model doesn't have us getting ripped apart at all, in fact—all the ordinary matter gets eaten up by black holes, which then evaporate away into nothing something like 10^{100} years from now."

"Oh, yeah, that's a *lot* better, dude. You should've quit while you were ahead."

"More importantly, though, that's not the end."

"I don't know about you, but getting eaten by a black hole sounds pretty final to me."

"OK, I'll give you that. What I meant was that this isn't the end of relativity."

"Dude, you've gone from moving cats to the end of the universe. What more can there be?"

"Well, there are still some really big mysteries remaining. General relativity is a fantastically successful theory, but it doesn't fit well with the *other* great theory of modern physics: quantum mechanics. That's the greatest outstanding mystery in physics, so the story isn't nearly over yet."

"Yeah? How's the rest of it go? No, wait, let me guess—that's the next chapter."

"Got it in one."

Chapter 12

THE UNIFIED THEORY OF CRITTERS: UNIFICATION OF FORCES

I'M PLAYING AROUND WITH THE COMPUTER, when I realize I haven't heard anything from the dog in a while. I go into the living room, where there are papers and crayons strewn about from when our daughter was coloring earlier. Emmy is standing over a piece of paper, with a green crayon in her mouth.

"Hey!" I say. "No eating the crayons! We've been over this."

"I wasn't eating it," she says. "I was using it to do physics!"

"Really. In crayon?"

"It's hard to use a pencil without thumbs. I can draw with the crayon in my mouth. I'm working on a Unified Theory of Critters!" She wags her tail proudly.

"Critter unification?"

"Yeah. See, my theory is that bunnies and squirrels are just different manifestations of a single unified critter. At a higher temperature, bunnies and squirrels would merge into a single type of thing."

"Bunnies and squirrels together at high temperature? That's Brunswick stew, isn't it?"

"Oooh! That's a good name. I was going to use 'Bunny-Squirrel Plasma,' but I like the way you think!"

"Anyway, that's not physics, it's cooking. Nobody does unified theories of bunnies and squirrels—they're about subatomic particles and fundamental forces. Unified theories don't work for biological organisms."

"Which brings up an interesting question."

"Yes?"

"How *do* unified theories work? Because, honestly, I have no idea."

"I thought you were working on a Unified Theory of Critters?"

"I said I was working on it. I didn't say I was getting anywhere. Right now, it's pretty much just an idea. But you know about physics stuff, so maybe you can help me!" She sits down, tail wagging, and tries to look attentive.

I sigh. "Well, you were on the right general track. The idea of a **unified theory** is to show that two forces that appear to be different things are actually the same thing. The simplest example is electromagnetism, which is the theory showing that electricity and magnetism are actually a single unified interaction."

"They are? That doesn't seem right. I mean, you have magnets on the fridge and electricity in lightbulbs, and those don't look like each other at all."

"Not in those forms, but if you run current through a coil of wire, you can make an electromagnet, which behaves like a magnet as long as the current is running."

"Oh, right. Humans use them to pick up cars and drop them on cats."

"I'm going to pretend I didn't hear that. You can also generate electric current by moving a magnet near a loop of wire. That's how we generate the electricity we use to power household appliances: running water or steam spins magnets, generating electrical current that we send through power lines."

"Which we use to keep our house warm and our food cold!"

"Right. Those phenomena show that electricity and magnetism are aspects of the same thing. The mathematical equations showing how they fit together were developed through the 1800s and collected together as Maxwell's equations."

"Oh, right. Maxwell's equations."

"Exactly. Physicists have also managed to unify electromagnetism with the weak nuclear force."

"Why's it called that?"

"Because it affects things in the nucleus. And, um, it's not as strong as the other nuclear force, which is—"

"Let me guess, the strong nuclear force."

"Yeah."

"Clever names, dude." She gives me a pitying look. "You should've stuck with Brunswick stew."

"Moving right along, there are theories about how to unify electromagnetism and the weak force with the strong force, but that's not understood nearly as well. And nobody has a good theory unifying those three with gravity."

"Why not?"

"Well, because it's a hard problem. Einstein spent the last twenty-odd years of his life working on it and got nowhere. Hundreds of physicists have worked on it in the fifty-plus years since he died and still don't have an answer."

"Maybe I can figure it out!"

"I doubt it. It involves a lot of math."

"Oh. I'm not so good with math." Her ears droop.

"I know."

"I guess that means I won't be writing a unified theory, then."

"Probably not." I scratch behind her ears, and she wags her tail.

"In that case," she says brightly, "I'll just have to eat these crayons!"

The concept of a unified theory of something or other is another idea that has escaped pure physics circles and entered the wider culture, at least as a sort of shorthand for a massive, all-encompassing intellectual goal that remains unattainable. Outside of geeky jokes, people studying the unification of fundamental forces now tend to call it something else—the name Grand Unified Theories refers to a set of failed theories from twenty to thirty years ago, while current attempts to pursue the same goal are typically referred to by a different label, even when the goal is the same.

The idea of unification is an appealingly simple one. It says that the array of different interactions we see in the universe are all, in fact, disguised aspects of a single interaction, in the same way that electricity and magnetism are different aspects of a single electromagnetic interaction. This would reduce the physics describing the universe from a vast tangle of equations and constants with seemingly arbitrary values to just one equation, ideally one simple enough, as a physics joke has it, to fit on a T-shirt.

This grand goal has consumed a huge amount of intellectual effort from theoretical physicists over the last century or so. The pursuit has yielded many fascinating discoveries, but a single equation for the universe remains almost as elusive as it was a hundred years ago. We'll close this book by looking briefly at some of the ways people have tried and failed to unify the forces of nature and talking about some of the prospects for the future.

MAKE ME ONE WITH EVERYTHING: UNIFICATION OF FORCES

Maxwell's equations, discussed back in Chapter 2, are the first and best example of the unification of forces. Decades of research into the apparently separate forces of electricity and magnetism in the early 1800s showed connections between them—current-carrying wires deflect the needles of nearby compasses, and moving magnets produce currents in nearby loops of wire—but Maxwell showed that all these different interactions could be pulled together in a single set of equations. What physicists had thought of as two separate forces were revealed to be complementary parts of a single interaction, the electromagnetic interaction.

The pursuit of unification in the modern sense began with Einstein, who in the 1920s started trying to find a way to unify the electromagnetic interaction with gravity. This seemed like a promising approach—the mathematical equations for the gravitational force between two masses and the electrostatic force between two charged particles look very similar—and Einstein's 1920 popular book *Relativity* even includes a note suggesting that such a unification was imminent.*

The quest for a unified theory consumed most of the last thirty years of Einstein's life, until his death in 1955, but never led to his hoped-for success. Along the way there were many interesting mathematical curiosities and discoveries, some of which have resurfaced in other contexts, but Einstein was ultimately thwarted by two factors: the existence of additional forces not known when he started his search and quantum mechanics.

The two additional forces are the blandly named the strong nuclear force and weak nuclear force, which operate only within the nuclei of atoms, over distances of 10^{-15}m or less. Their properties couldn't be adequately investigated until the particle accelerator era started in the 1950s, but they are now central to our understanding of the nature of matter. The strong nuclear force binds quarks into protons and neutrons and holds protons and neutrons together in the nuclei of atoms. The weak nuclear force transforms particles of one type into another, which is critical for both radioactive decay and the fusion reactions that power the sun.

The other main obstacle to Einstein's quest for unification was quantum mechanics. Ironically, Einstein had a hand in launching the theory—his Nobel Prize citation in 1921 does not mention relativity but specifically cites his quantum theory of the photoelectric effect—but as it developed,

---

* A footnote in Chapter 16 reads, "The general theory of relativity renders it likely that the electrical masses of an electron are held together by gravitational forces."

he found it philosophically distasteful, so he never fully accepted it. He always hoped for a deeper theory that would replace the disturbing elements of quantum physics with laws in a more traditional mode.

We know now, however, that quantum mechanics, even the parts that Einstein found distasteful, is unequivocally correct in its description of nature. Quantum electrodynamics (QED), which combines quantum mechanics with special relativity, has been tested in the laboratory and agrees with experiments to fourteen decimal places, making it arguably the best-tested theory in the history of science. It provides the framework we use for describing electromagnetic, weak, and strong interactions, and it is an essential piece of the Standard Model of particle physics. No theory that hopes to describe all of reality can avoid incorporating quantum ideas, no matter how much Einstein may have wished otherwise.

"Wait, he didn't believe in the best-tested theory in the history of science, but you call the cosmological constant his greatest blunder? Isn't not believing in quantum physics way more blunderous?"

"*I* didn't call the cosmological constant his greatest blunder; *he* did. And he wouldn't've admitted that not believing in quantum mechanics was a blunder because he thought he was right and quantum theory was incomplete."

"But you said it's the best-tested theory in the history of science!"

"It is now, but it wasn't in 1955, when Einstein died. During Einstein's lifetime, there was always room for possible theories beyond quantum physics, and most of the weird effects that bothered Einstein hadn't been directly observed yet. Only after decades of further research can we now definitively say that the universe really is as weird as quantum mechanics predicts."

"Yeah, but think how much further we could've gotten if he hadn't been so deluded!"

"Maybe. Then again, many cool quantum experiments were directly inspired by Einstein's attacks on quantum theory, so it wasn't all negative. And anyway, he might not have been able to solve the unification problem even if he had accepted quantum mechanics."

"Why not?"

"Well, because it's really difficult. The standard techniques for dealing with quantum forces break down when you try to apply them to gravity."

"Why is that?"

"Well, first, I need to explain the standard techniques, OK?"

"OK."

PUSH ME, PULL YOU: EXCHANGE FORCES

As we discussed in Chapter 8, the central idea of QED is that forces are mediated by **exchange bosons**. This provides the basic paradigm used to describe all interactions in quantum mechanical terms. The electromagnetic force is carried by photons; the weak nuclear force is carried by the W^+, W^-, and Z bosons; and the strong nuclear force is carried by gluons, which come in eight different varieties.*

The exchange-force picture of forces works extremely well for the electromagnetic and weak forces, which were unified into the electroweak force by Steven Weinberg, Abdus Salam, and Sheldon Glashow during the 1970s. The mathematical theory of strong-force interactions also appears to fit the exchange-force paradigm. The strong-force picture is complicated by the fact that, unlike photons, gluons interact with one another, greatly increasing the number of diagrams and the complexity of calculations. The basic picture is the same, though: the strong force is communicated by the exchange of particles, and calculations involving the strong force use Feynman diagrams.

"Isn't that a little vague, dude?"

"Maybe a tiny bit, but I'm trying to skip over the details because the strong force is really complicated."

---

* Eight, because the strong-force analogue of electrical charge comes in three varieties, called "colors," and each type of gluon communicates the force between two different colors. The quark "colors" are the equivalent of positive and negative charges in electromagnetism: quarks with different colors are attracted to one another. The subatomic particles we directly detect are "colorless" combinations of quarks, either groups of three quarks, one of each color, or a quark and an antiquark with one color and its anticolor.

"Complicated, how?"

"For example, the nature of the strong force is that you never see quarks and gluons in isolation. Unlike the other forces, where the strength of the interaction decreases as you pull things farther apart, the strong interaction gets even stronger as the distance increases. When you try to pull two quarks apart, the energy associated with the strong force becomes great enough to create new quark-antiquark pairs, so you end up with several multiquark particles but no free quarks."

"So there's no way to ever see a free quark or gluon?"

"Not really. The closest you can get is the 'quark-gluon plasma' created in high-energy collisions of gold atoms at the Relativistic Heavy Ion Collider (RHIC) on Long Island or of lead atoms at the Large Hadron Collider (LHC)."

"I thought you said there wasn't any point in colliding heavy and complicated things?"

"I said you couldn't simply scale up the collision energy by using heavier particles. But that's not what RHIC and the LHC are doing. Their whole goal is to create a system where you have a huge number of high-energy particles running around. If the energy is high enough, you can break the quarks out of the protons and neutrons inside the nuclei and into a plasma of lots of quarks, plus innumerable gluons flying around conveying the strong force. This quark-gluon plasma is a condition that existed in the first seconds after the Big Bang, and studying its properties will tell us a lot about the behavior of the strong force and how it fits with the others."

"So what's the conclusion?"

"It's still up in the air. The heavy-ion experiments only started about ten years ago, and they only confirmed the creation of a quark-gluon plasma five years ago. It's one of the hot topics in physics at the moment, though, and with the LHC coming online, they'll have even more data to study. So the situation is very fluid."

"You know what? You were right."

"Right about what?"

"That is awfully complicated."

"But the basic paradigm is the same: the strong force is carried by exchange particles called gluons, whose properties are fairly well constrained."

"So, is there anything that departs from that paradigm?"

"Yes, and we've already talked about it."

"We have? Oh, gravity!"

"Exactly."

ONE OF THESE THINGS IS NOT LIKE THE OTHERS: THE PROBLEM OF QUANTUM GRAVITY

Conspicuously absent from the list of exchange forces and particles above is any mention of gravity. There is a name for the hypothetical particle exchanged to carry gravity, the **graviton**, but not much beyond that. For the moment, at least, gravity doesn't comfortably fit the same paradigm as the other fundamental forces.

When we say that QED is arguably the most precisely tested theory in the history of science, the only other theory in the running is general relativity, which has also passed an astounding range of tests. Einstein's theory has held up brilliantly, making correct predictions on scales from small laboratory experiments all the way up to galaxies and clusters of galaxies—which is both good and bad: on the good side, it represents a towering intellectual achievement; on the bad side, it doesn't work well with quantum mechanics, even though both theories are spectacularly successful on their own.

There are two ways to understand the problem with quantum gravity, both of which involve the particles that turn up in Feynman diagrams. Coming at the problem from the quantum mechanics side, if you simply declare that gravity is carried by gravitons and use standard QED techniques to calculate the energy of anything, every calculation gives you the same answer: infinity.

The problem is that something like an electron is a point-like particle* with a nonzero mass. This requires the electron to be infinitely dense, in

* The electron is known to be smaller than 10^{-22}m in radius, one-trillionth the size of an atom. In the Standard Model, it is believed to be a true point, with no measurable size.

some sense, and an infinitely dense mass has a large gravitational energy, which means lots of gravitons. But gravitons themselves carry energy, thus have mass, which means that they interact with each other, producing more gravitons, and so on. Matters quickly get out of hand, leading to infinite energy for any particle, even a graviton by itself.

QED has a similar problem, as electrons also pack a finite electric charge into infinitesimal space and thus seem at first to have an infinite energy. This problem is fixed in part by the appearance of virtual particles: the space around the electron is full of virtual particle–antiparticle pairs popping into and out of existence, and some of those particles are positive. The presence—even fleetingly—of positively charged particles in the space around the electron effectively "spreads out" the charge, avoiding the infinities you would otherwise encounter in calculating its energy.

The same trick can't rescue a quantum theory of gravity, though, because any particle carrying energy interacts with gravitons in the same way. Unlike in the QED case, more complicated processes involving many virtual particles end up making *more* important contributions to the final energy. The standard techniques aren't enough to avoid getting infinite results.

Coming at the problem from the direction of general relativity doesn't fare any better. The central idea of general relativity is that matter curves spacetime, but quantum mechanics tells us that even empty space is full of particles popping in and out of existence. Each of those particles must bend spacetime, which means that on a quantum level, the structure of spacetime must be constantly and randomly fluctuating, taking on all possible values at any given time. This is nearly impossible to reconcile with the usual tools of general relativity, which work on a smooth and continuous spacetime with a well-defined curvature at any point.

However you attack it, the mathematical techniques that produce such fabulous successes when applied to quantum physics break down when applied directly to the problem of gravity. In their own realms, quantum mechanics and general relativity are each fantastically successful, but together, they're like—

"If you say they're like cats and dogs, dude, I swear I'm going to bite you."

"What?"

"It's sooooo cliché, dude. I mean, really, after 291 pages, do you really want to sink to that level of hackwork?"

"Good point."

"Anyway, I thought people already had combined quantum mechanics and general relativity? Isn't that what Hawking radiation is about: putting quantum mechanics together with general relativity in a black hole?"

"Only in the most rudimentary way. The theory that gives us Hawking radiation is a patch job that barely uses general relativity at all, treating the curvature of spacetime as fixed and unchanging. The virtual particle production and radiation goes on against that background. You can get away with that, because the typical masses and distances involved are relatively large, in the same way that you don't need to worry about the wave nature of matter to figure out Earth's orbit, because its quantum wavelength is so absurdly short."

"Sooo . . . where do you need quantum gravity then?"

"When you're talking about really large masses and really short distances. The singularity at the center of a black hole, for example, would need a theory of quantum gravity to describe it."

"But that's inside the event horizon, right?"

"Well, yeah."

"If the only thing you need it to describe is inside a horizon that nothing, not even light, can escape, who cares?"

"Well, quantum gravity's not likely to produce a flying car, if that's what you're after—"

"Darn! A flying car would be awesome!"

"—but it would also be needed to describe the entire universe in the first infinitesimal fractions of a second after the Big Bang. And if we understood that, we would know something about exactly how the universe began, and big questions don't come any bigger than that."

"OK, that's a good point. I'd still like a flying car, though. Think of all the birds we could chase!"

STRINGS AND SEALING WAX AND OTHER FANCY STUFF: PATHS TO UNIFICATION

Reconciling quantum mechanics and general relativity is the deepest and most fundamental problem facing theoretical physics today. Numerous approaches have been tried, but all have fallen short, either by predicting particles or effects that don't show up or by failing to explain phenomena that *are* observed. Theoretical physicists have been forced to adopt increasingly elaborate methods in order to make the two greatest theories in physics fit together.

String theory is the most popular current candidate for a unified theory of quantum gravity that would combine gravity, electromagnetism, and the nuclear forces into a single unified mathematical framework. It originated in the 1960s with the realization that the mathematical equations for some particle interactions, written in the right way, are identical to the equations describing a vibrating string. In this picture, all fundamental particles—both the material particles (quarks and leptons) and the force-carrying bosons—are described as short lengths of "string." The different particle types are distinguished by their vibrations, so the same bit of string can be either an up quark or a top quark, depending on how it vibrates.

The idea of string theory is conceptually and mathematically appealing, and as other models have failed one after another, it has consumed a greater share of the efforts of theoretical physicists. It is not without its problems, though, among them that the string description only works if the strings are free to vibrate in ten or eleven dimensions of spacetime, instead of the familiar four.

"So, these strings exist in parallel universes?"

"No, these aren't **extra dimensions** in the science fiction sense of parallel worlds where Mr. Spock is evil and has a beard. These are additional directions of motion, beyond the usual left-right, forward-back, and up-down that we're used to."

"So, what, diagonally?"

"No, an entirely new direction, at right angles to the usual directions, in the same way that up is at right angles to left."

"Oh, I see . . . Actually, no I don't. That just makes my head hurt."

"It's not easy to picture, because we only deal with three dimensions of space in our normal lives."

"Yeah, but doesn't that constitute a huge problem for the theory right there? I mean, if the universe had eleven dimensions, don't you think we'd notice them?"

"It's a problem, all right, but not an insurmountable one. The extra dimensions could be there, but so small that we can't see them."

"What does that even mean?"

"It just means that these dimensions might be bounded in some way, like a narrow dog run, so you can't move very far in that dimension. Or they might be curled up into loops, so that if you try to go more than a tiny distance, you end up back where you started. We might have extra dimensions of space, but because we can only move a tiny fraction of a millimeter in them, we don't notice them in everyday life."

"So, if they're too small to see, how is that different from their not existing at all?"

"Well, single electrons are too small to see, right?"

"Sure."

"But they can have measurable consequences in physics experiments. Similarly, the extra dimensions in string theory might also have some consequences for high-energy physics experiments, even though they're too small to see in everyday life."

"I guess so, but it still sounds awfully baroque."

"Oh, it is."

String theory has great promise as a theory to unify gravity with the other forces, but the presence of hidden extra dimensions complicates the theory enormously. There's only one way to "curl up" a single extra dimension, but as you add more extra dimensions, the ways of hiding them multiply dramatically. Putting the large number of these arrangements together with the large number of ways strings can behave gives you an effectively infinite number of possible versions of string theory.* The complexity of the mathematics involved also makes it extremely difficult to make concrete predictions about experiments.

String theory research has proven very mathematically fruitful, however, and techniques derived from it have found useful applications in surprising places: one particular technique for studying strongly interacting particles has been used to describe the quark-gluon plasma observed at RHIC and the LHC, as well as the behavior of electrons inside some solid materials. Theorists hope that continued mathematical work will provide further insights and that the results expected from the Large Hadron Collider and other experiments will inspire some future breakthrough.

"How does smashing protons together help?"

"Well, for one thing, the complexity of string theory means that it predicts many particles beyond the Standard Model ones that we know and love. The LHC might detect some of these particles."

"But that doesn't tell you about extra dimensions, does it?"

"Indirectly it does, as their properties would tell you something about the ways strings vibrate, which is constrained by the geometry of the extra dimensions. There's also a chance they could be observed more directly, through what you see in the LHC. Or, rather, what you don't see."

"What?"

"Well, a lot of string theories have most of the particles we normally deal with stuck in our three-dimensional space, like dogs and bunnies trapped near the ground by gravity. Gravitons, however, can escape into the extra dimensions, like birds that can fly out of the yard. If this is true, particle collisions will occasionally produce gravitons, which would then escape, taking some energy with them. This would show up as missing

---

* There have been several different attempts to "count" the number of possible string theories, and numbers like 10^{300} and 10^{500} get bandied about. Those are so large as to have no meaning for most human scientists, let alone dogs.

energy that can't be accounted for by the mass and motion of the particles that are detected."

"So, you smash things together, and when you lose energy, you say it went into the extra dimensions?"

"More or less. This is a really tricky problem, though, because there are lots of ways to create particles that don't get detected—ordinary neutrinos, for example. It would require sophisticated statistical analysis to confidently attribute the missing energy to escaping gravitons."

"I don't think I'd believe that, dude. Can't you do anything more direct?"

"Well, there's always the chance of forming a black hole. For that to happen at the LHC, extra dimensions would pretty much be required."

"What do black holes have to do with extra dimensions?"

"Well, if gravitons really do escape into extra dimensions, then the force of gravity that we see as really weak would actually be much stronger in the full universe."

"Gravity is weak? How do you figure? If anything, gravity is too strong—it keeps me stuck to the ground all the time. If gravity were weak, I'd be unstoppable."

"Yes, but we feel significant gravitational forces only because the mass of the Earth is gigantic—the gravitational force between you and me, for example, is way too small to notice. You can see how ridiculously weak gravity is from the trick where I rub a balloon on your fur, then stick it to the ceiling. The electromagnetic force of one extra electron for every trillion or so atoms in the balloon easily overcomes the gravitational force of the entire Earth pulling the balloon down."

"OK, but how do extra dimensions help this?"

"Well, if gravity is weak because gravitons are escaping into extra dimensions, you should be able to catch its full strength by getting closer than the size of the extra dimensions. To make a black hole at the LHC using the ordinary three dimensions requires you to pack all the energy of the collision into a space just 10^{-50}m across, which is basically impossible. If there are largish extra dimensions, though, when you start packing things really close together, gravity should get dramatically stronger at distances comparable to the size of the dimension."

"Because if two masses are separated by a small amount, the gravitons from one get snagged by the other before they can escape into the extra dimensions?"

"That's the idea, yes. And if gravity does strengthen at short distances, you might be able to make a black hole by packing the entire energy of two colliding protons into a more manageable 10^{-18}m or so. That's the only way a black hole could be formed at the LHC, and if it did happen, it would be spectacular evidence that extra dimensions are real."

"So, you're saying that string theorists are rooting for a black hole that will eat the Earth?"

"Remember, it wouldn't really eat the Earth; it would evaporate almost instantly. And I wouldn't say they're rooting for it, exactly—it's too remote a possibility—but if the LHC did see an evaporating black hole, there would be much rejoicing in some quarters of the theoretical physics community."

"Humans are so weird. I mean, you have a theory that requires the existence of extra dimensions nobody has seen, predicts particles nobody has seen, and might make it possible to create black holes, but probably not. And that's the best candidate for a theory of quantum gravity?"

"Pretty much. It tells you just how difficult this is."

"It tells me that you all need to go outside and chase some bunnies, that's what it tells me."

"Yeah, well, you may be right there."

String theory is the most popular of the candidates for a theory of everything, offering as it does both a quantum theory of gravity and a theory of all the other forces of nature. The most notable other approach to combining general relativity and quantum mechanics is probably **loop quantum gravity,** which quantizes spacetime itself. In this approach, there is a minimum possible spatial distance and a minimum possible time interval,* with spacetime itself put together from little "loops" of the appropriate size, like a knitted dog blanket. This method has produced some successes

* Around 10^{-35}m and 10^{-44}s, respectively.

but does not unify gravity with the other forces, seeking only a quantum theory of gravity to which other interactions must be added.

Whatever approach you prefer, the problem is an extraordinarily difficult one, and in many ways, we're not much closer to a solution now than we were in Einstein's day. Generations of theoretical physicists have spent their careers convinced that we were on the verge of explaining absolutely everything, only to have the ultimate prize slip from their grasp.

Although, in many ways, it's a frustrating time to be a theoretical physicist—some of this frustration boiled over a few years back in a spate of anti–string theory books—it's also an exciting time. While exploration of general relativity has produced almost as many questions as answers, telling us that 96 percent of the universe is in forms we don't currently understand and frustrating all attempts to combine general relativity with quantum mechanics, the steady improvement of technology over the last few decades has put us in position to start getting answers. New generations of experiments—both giant projects, like the Laser Interferometer Gravity-Wave Observatory and the LHC, and tabletop experiments looking for new physics on more ordinary scales—are poised to produce results in the next few years. We can't say for sure what these experiments will find—the safest bet is probably that they'll find something other than what we expect—but whatever they find will be a test of existing theories and the basis for new theories. There's nothing more exciting to a scientist than new facts about the nature of the universe, so the next several years should be very interesting.

"Not even bunnies?"

"Pardon?"

"You said there's nothing more exciting than new facts, but bunnies are way more exciting than facts."

"To a dog, maybe. I'm not fast enough to chase bunnies very effectively, though, so they don't do much for me. When I want food, I go to the store."

"Good point. If I could buy bacon myself, I might not chase bunnies so much. Then again, I might chase them more. Chasing bunnies is fun."

"Whatever makes you happy."

"Anyway, this relativity stuff was pretty cool. It kind of makes my head hurt, but it's neat all the same. And it's good to check off another area of physics."

"Check off?"

"Yeah. I mean, now we've talked about everything there is to talk about in relativity, right?"

"Hardly. There's tons of cool stuff we didn't get to. We barely touched on gravitational waves and didn't talk about wormholes at all."

"Wormholes?"

"If you warp space in just the right way, you can arrange for two very distant points to be connected by a sort of tunnel between them."

"So they're a shortcut to distant stars?"

"Potentially. They also give you a way of making a sort of time machine, where it's much later on one end of the wormhole than the other."

"That'd be pretty neat."

"Yeah. There are also even more exotic theories of the universe and its history and about other universes that might exist, and so on. What we've talked about in this book really only scratches the surface of what general relativity has to offer. I can recommend a bunch of other books if you want to know more."

"Sure, only . . ."

"Yes?"

"Can we go for a walk and chase some bunnies first?"

"Sure, we can do that."

"Woo-hoo!"

ACKNOWLEDGMENTS

As with any book, a huge number of people made the writing of this one possible. I'm particularly grateful to a bunch of fellow physicists and astronomers who patiently answered my silly questions and corrected my mistakes: Aaron Bergman, Sean Carroll, Tommaso Dorigo, Becky Koopmann, Jon Marr, Moshe Rozali, and Jason Slaunwhite. Any remaining misstatements are entirely my own doing, and no reflection on them.

This book started life as a set of posts on my weblog, *Uncertain Principles* (http://scienceblogs.com/principles/), and thanks are due to the people who have kept ScienceBlogs running these last several years: Christopher Mims, Katherine Sharpe, Arikia Millikan, Erin Johnson, Evan Lerner, and Wes Dodson. Thanks, too, to my readers and commenters, who keep the whole business interesting.

This book would not exist without my agent, Erin Hosier, and my editor, TJ Kelleher. Thanks for giving me the opportunity to spend some more time talking to the dog about physics and sharing that with other people.

I've been very lucky to know a lot of great dogs over the years—Patches, Rory, RD, and even Tinker, among others. Because relativity is all about different observers, I had the opportunity to include a few of them here: Bodie is my parents' yellow Lab, Truman was my in-laws' Boston terrier, Maeve is Patti and Steve Severson's Springer Spaniel, and Winthrop is the grand old basset hound of Niskayuna, belonging to Mike Pletman. Special thanks to Pam Korda and Sarah Nichols, whose generous donations to the educational charity DonorsChoose (http://DonorsChoose.org) earned mentions for Harley the poodle and Anson the Lab mix. They're all excellent dogs, and deserve extra treats. And thank you to Claire Foster for the drawings of all the various animals that appear in the figures.

We adopted Emmy from the Mohawk & Hudson River Humane Society shelter in Menands, New York (http://mohawkhumanesociety.org). Like most such organizations, they are a terrific source for wonderful dogs (and even some pretty OK cats), and if you're looking for a new pet, Emmy and I encourage you to look at a shelter in your area. Thanks also to the nice folks at Sunnyside Veterinary for keeping Emmy healthy and happy, and to Gaye, Camille, and the others at I'm a Pet Nanny pet-sitting service for keeping Emmy company when her humans go out of town.

Surprisingly, writing a book with a toddler in the house turns out to be something of a challenge. As a result, large parts of this were written at Barnes and Noble, Starbucks, and Panera in Niskayuna, and the Pasta Factory in Latham. Thanks for keeping me supplied with food, drink, and Wi-Fi while I wrote.

Huge thanks go to my parents and Kate's, for all their support and occasional babysitting. I wouldn't have made it through this process with my sanity intact without their help. And, of course, thanks most of all to my wife, Kate Nepveu: for putting up with all the inconveniences of writing; for listening to me kicking ideas around and ranting about obscure problems; for gently correcting my grammar and making me look much smarter than I would otherwise; for Claire and David, who have made my life indescribably richer; and most of all for making the whole thing possible by laughing at my silly jokes. This would literally never have happened without her.

GLOSSARY

acceleration: The rate at which an object changes its velocity, either by changing speed, as when a dog starts chasing a bunny, or by changing direction, as when a dog chases her tail in a circle. Newton's laws relate the acceleration to the force applied, and the equivalence principle says that the effects of acceleration are indistinguishable from those of gravity.

antimatter: Matter composed of "antiparticles." Every particle has an antimatter equivalent, with the same mass but the opposite electrical charge. When a particle comes in contact with its antiparticle, they annihilate each other, converting all their mass into energy. The visible universe consists entirely of ordinary matter, but antimatter is created by cosmic rays, radioactive decay, and collisions in particle accelerators.

atomic clock: A type of clock whose rate of "ticking" is the frequency of the oscillating electromagnetic field of one of the characteristic colors of light emitted by a particular atom. Quantum physics guarantees that all atoms of any given element are identical, making atomic clocks the most precise clocks possible.

Big Bang cosmology: Our best current understanding of the history of the universe, which began as an extremely small, hot, and dense point about 13.7 billion years ago and has been expanding and cooling ever since.

black hole: An object whose mass is concentrated into a region smaller than the Schwarzchild radius for that mass. The curvature of spacetime produced by a black hole is so extreme that not even light can escape from the region inside the event horizon, making it an excellent place to store cats. Black holes are produced by the collapse of very massive stars and evaporate slowly through the emission of Hawking radiation.

black hole information paradox: An outstanding puzzle for theories of physics that regard information as fundamental. To preserve the information content of the universe, the Hawking radiation from an evaporating black hole must somehow contain all the information about the objects that fell into the black hole, but it is not clear how.

causality: The idea that effects must always be preceded by their causes. So, for example, a dog will only start barking after receiving a signal (a light, sound, smell) indicating the presence of a threat to the house (a cat, the mail carrier, other dogs barking several miles away). The requirement of causality forbids faster-than-light travel or even signals, as they create situations where some observers see effects happen before causes.

conservation of energy: The principle that the total energy of a group of interacting objects must always add up to the same total value, though the energy can change from one form to another.

conservation of momentum: The principle that the total momentum of a group of interacting objects must always add up to the same value, though momentum can shift from one object to another.

conserved quantities: Quantities like energy or momentum whose total value does not change over time. According to Noether's theorem, conserved quantities are associated with symmetries in the laws of physics, so, for example, momentum is conserved because the laws of physics do not depend on position in space.

cosmic microwave background: Radiation left over from the very early universe, dating to about three hundred thousand years after the Big Bang, when the temperature was extremely high. As the universe has expanded, it has also cooled, so this radiation is now found in the microwave region of the spectrum, with a temperature of 2.7K.

cosmic rays: Energetic particles from outer space, mostly protons, thousands of which enter every square meter of the Earth's atmosphere every second. Cosmic rays were used for early particle physics experiments, but the predictability of particle accelerators has made these machines the dominant tool for particle physics since the 1960s.

cosmological constant: An energy associated with empty space that causes spacetime to expand. This constant was initially introduced by Einstein in order to counter the gravitational attraction of matter and produce a universe whose size remained constant. When Edwin Hubble discovered that the universe was expanding in time, Einstein renounced the cosmological constant, calling it his "greatest blunder." Since the late 1990s, though, astronomers have determined that the expansion of the universe is accelerating, which they attribute to a cosmological constant.

cosmology: The study of the universe as a whole, its origin, history, and eventual fate.

critical density: In cosmology, the "just right" amount of energy in the universe to make the overall curvature of the universe zero and produce an expanding universe whose rate of expansion slows down over time.

curvature: In cosmology, the parameter describing the four-dimensional shape of spacetime due to the matter and energy in the universe. If the universe contains more than the critical density of matter and energy, it has positive curvature, like the surface a sphere. If the universe contains less than the critical density of matter and energy, it will have negative curvature, like the surface of a saddle. If the universe contains matter and energy at exactly the critical density, it will have zero curvature, and the familiar rules of geometry on flat surfaces will apply.

dark energy: An energy associated with empty space that drives the accelerating expansion of the universe. Its exact source and nature is still unknown. Dark energy accounts for about 73 percent of the total energy content of the universe.

dark matter: An unknown type of matter found in galaxies and galaxy clusters, so called because it interacts gravitationally like ordinary matter but does not emit any light. Also, astronomers, like physicists, are terrible at naming things. Dark matter accounts for about 23 percent of the total energy content of the universe.

Doppler effect: A shift in the frequency of waves emitted by a moving source. The frequency shifts up when the source is approaching and down when the source is receding. The best-known example involves sound waves: the horn on a speeding car will shift up in pitch as it approaches a dog on the side of the road and down as the car moves past. The Doppler effect also changes the frequency of light, with light from approaching sources shifted toward the blue end of the visible spectrum, and light from receding sources shifted toward the red.

electromagnetic waves: Maxwell's equations explain light as an electromagnetic wave composed of an oscillating electric field and an oscillating magnetic field sustaining each other as they move through space. These waves move at a single speed that does not depend on the speed of the source or observer, a prediction that conflicts with classical ideas and led directly to the development of relativity.

electron: One of the particles making up ordinary atoms; they have small mass and a negative charge. Electrons are the most familiar type of lepton.

embedding diagram: A picture illustrating the distortion of spacetime by matter in general relativity by stretching a two-dimensional slice of space in a fictitious third dimension, as if it were a rubber sheet with masses placed on it.

energy: A quantity associated with the motion of an object or its ability to start moving. A running dog has kinetic energy, while a dog at rest has potential energy, because at any moment she might start running around like a mad thing. The total energy of a collection of interacting objects is conserved, which Noether's theorem tells us happens because of symmetry in time: because the laws of physics work the same in the future as in the past, there must be a physical quantity that does not change as time passes, which is the energy.

equivalence principle: A principle stating that the effects of gravity are indistinguishable from the effects of acceleration. It was introduced in a weak form in the 1600s by Galileo, who showed that all objects fall to the surface of the Earth at exactly the same rate regardless of their composition. Einstein realized in 1907 that this principle let him expand special relativity into a theory of accelerated motion and gravity, which led to the theory of general relativity in 1915.

event horizon: The boundary between the interior of a black hole and the rest of the universe. Anything that crosses the event horizon into a black hole—interstellar gas, annoying space cats, even light—can never return.

exchange bosons: Fundamental particles in the Standard Model that pass back and forth between other particles to convey the fundamental interactions of physics. The electromagnetic force is carried by photons;

the weak nuclear force is carried by the W^+, W^-, and Z bosons; and the strong nuclear force is carried by twelve varieties of gluons.

extra dimensions: Directions of motion through space at right angles to the three usual dimensions (up-down, left-right, forward-back). Trying to picture this gives most humans, let alone dogs, a headache, but string theory requires several such dimensions, which may be "curled up," making them too small to detect.

Feynman diagram: A calculational tool devised by Richard Feynman in his formulation of quantum electrodynamics. Each diagram describes a possible sequence of events in the history of a group of interacting particles, involving the exchange of force-carrier bosons and other virtual particles. Each diagram also represents a mathematical calculation; to determine the energy of a particle, you draw all the diagrams that might contribute to its energy and add the results of the calculations.

frame of reference: The measurements of the position and time of various events made by a particular observer. Relativity provides a framework for converting between different frames of reference. So, for example, if you know the position and time measured by a stationary dog, you can use them to determine the position and time measured by a moving cat, and vice versa.

general relativity: The full theory of relativity completed by Einstein in 1915, which extends his 1905 special theory of relativity to cover all possible moving observers and explains gravity as the bending of spacetime by matter.

Global Positioning System (GPS): A network of atomic clocks in satellites broadcasting time signals. Measuring the arrival times of signals from different satellites fixes the position of the receiver on the Earth to within a few meters, provided the effects of relativity on the orbiting clocks are taken into account. GPS is an important navigational aid for humans, whose pathetic little noses aren't good enough to sniff out the right way home.

gluons: The exchange bosons that carry the strong nuclear interaction between quarks. Gluons come in twelve different varieties, depending on the "color" of the interacting quarks.

gravitational lensing: The bending of light by gravity, predicted by general relativity, which can produce multiple distorted images of a distant galaxy due to a massive object between Earth and the target galaxy.

Astronomers can use the lensed images to determine the total mass of, and distribution of mass in, the intermediate object.

gravitational redshift: The shifting of a beam of light sent upward in a gravitational field to a lower frequency (toward the red end of the visible spectrum), according to an observer at higher altitude. There is a corresponding blueshift for light sent downward. This effect has been confirmed experimentally using sources and detectors on a 20m tower.

gravitational time dilation: The slower ticking of clocks closer to a massive object than clocks farther away from the object, according to general relativity. This has been confirmed through the behavior of the atomic clocks in GPS satellites and directly observed for a 33 cm change in altitude using ultraprecise aluminum ion clocks.

gravitational waves: A stretching and compressing of spacetime itself that travels through the universe like a wave in a pond. When a gravitational wave passes, objects along the path of the wave will get infinitesimally farther apart, then closer together. The indirect effects of gravitational waves have been observed in binary pulsars, but they have not been directly detected yet, though several experiments are looking for them.

graviton: The hypothetical exchange boson carrying the force of gravity between interacting objects. The exact properties of the graviton are not known, because there is no working theory of quantum gravity yet; this is one of the great outstanding problems in theoretical physics.

gravity: An attractive force between two objects having mass, which general relativity tells us is caused by the curvature of spacetime. If not for gravity, dogs would be unstoppable.

Hawking radiation: Radiation emitted from the event horizon of a black hole due to virtual particles created near the event horizon. When one particle from a virtual particle–antiparticle pair falls into the black hole, the other escapes as Hawking radiation, and the black hole loses a corresponding amount of mass. Given time, the black hole evaporates entirely into this radiation.

Higgs boson: A hypothetical particle whose interaction with the other Standard Model particles is responsible for giving them their different masses; at least six different people came up with the same idea, though for some reason only Peter Higgs's name has been attached to it. As of December 2011, the Higgs boson has not been directly detected, though particle physicists are getting close.

inertial frame: A frame of reference in which Newton's laws of motion appear to hold. A dog at rest and a cat slinking by at constant speed in a straight line will each be in their own inertial frames of reference. Special relativity requires all of the laws of physics to work exactly the same way in any inertial frame.

invariant quantities: Physical quantities that do not change when you move from one frame of reference to another. In classical physics, the length of an object would be invariant, but in relativity, we find that the length of a fast-moving object decreases. The spacetime interval between two events, however, is the same for all observers.

kinetic energy: Energy associated with an object moving through space. In special relativity, the kinetic energy depends on the mass, the Lorentz factor (γ), and the speed of light squared.

length contraction: In special relativity, a moving object shrinks along the direction of motion, so a stationary dog watching a cat move past will measure it to be slightly shorter than when it is stationary, while the cat will see the dog as shorter. This effect was predicted in the 1880s but not widely accepted until Einstein explained it through the relativity of simultaneity.

lepton: A class of subatomic particles consisting of the electron, its heavier cousins the muon and tau lepton, and the electron, muon, and tau neutrinos.

light clock: A thought experiment to illustrate time dilation in special relativity, using a clock based on a pulse of light bouncing back and forth between two mirrors. An observer looking at a moving clock will see a longer time between "ticks" than for a stationary clock because the light in the moving clock follows a longer path at the same speed of light.

light cone: The region in a spacetime diagram between the worldlines of two light rays sent out in opposite directions. Events within the light cone of the central event can be connected to it by a chain of cause and effect; events outside the light cone of the central event can never cause, or be caused by, that event.

loop quantum gravity: A possible theory of quantum gravity in which spacetime is made up of "loops" of some minimum spatial size (around 10^{-35}m) and duration in time (around 10^{-44}s).

Lorentz factor: A numerical factor that appears again and again in relativity, given by

$$\gamma = \frac{1}{\sqrt{1 - \frac{v^2}{c^2}}}$$

where v is the speed of the object, and c is the speed of light. The Lorentz factor is usually represented by a lowercase Greek letter gamma (γ), because physicists are almost as lazy as dogs and don't want to write the full formula all the time.

luminiferous aether: The medium that was believed to support light waves the way air supports sound waves. Numerous experiments failed to detect the aether or the Earth's motion through it, and Einstein's theory of special relativity showed that it was superfluous.

Maxwell's equations: Four equations that provide a complete description of electric and magnetic fields and explain light as an electromagnetic wave. As a result, these equations appear on coffee cups and bumper stickers and other nerdy accessories. Maxwell's equations predict a single speed of light that does not depend on the speed of the source or the observer. This appears to conflict with common sense and precipitated a crisis that led to relativity.

Michelson interferometer: A measuring device using the addition of light waves to detect tiny changes in the speed of light. Invented by Albert Michelson for the Michelson-Morley experiment, Michelson interferometers are still used today to measure tiny changes in position in detectors for gravitational waves.

Michelson-Morley experiment: An experiment looking for a change in the speed of light due to the Earth's motion through the aether. Its failure to detect any change was a crucial step toward the development of relativity and the discarding of the aether theory.

momentum: A quantity associated with a moving object; it is the product of the Lorentz factor (γ), the mass (m), and the velocity (v). The total momentum of a collection of interacting objects is a conserved quantity and does not change, though momentum may be redistributed among the objects. Noether's theorem tells us that momentum conservation is a result of symmetry in space: because the laws of physics are the same for a dog in Schenectady and an elephant in Tanzania, there must be some quantity associated with moving objects that remains constant as you move from one place to the other, which is the momentum.

muon: A lepton, similar to an electron but with two hundred times the mass and a lifetime of around 2.2 μs (at rest).

neutrino: Neutral leptons with extremely tiny masses that interact with ordinary matter only through the weak nuclear interaction and thus are extremely difficult to detect. They play an essential role in many nuclear reactions, though.

neutron: A neutral particle found in the nuclei of atoms, consisting of two down quarks and one up quark held together by the strong nuclear interaction.

neutron star: The remnant of an extremely massive star, left behind after a supernova. If the collapsing core of the star has a mass greater than 1.4 times the mass of the sun, the inward force of gravity compresses protons and electrons in the star until they combine to form neutrons. These neutrons resist being packed more tightly with enough force to halt the collapse, provided the mass is not too large. If the core mass is greater than about three times the mass of the sun, even this neutron pressure can't withstand gravity, and the star forms a black hole.

Newton's laws: Isaac Newton laid out three laws governing the motion of ordinary objects in his *Philosophiae Naturalis Principia Mathematica,* which marked the birth of physics as a mathematical science. Newton's first law says that an object does not change its motion unless an external force makes it change—so, for example, a sleeping dog tends to remain at rest until her human wakes her, while a running dog will keep on running at constant speed until her leash makes her stop. Newton's second law quantifies the effect of a force, explaining how long and hard a dog needs to pull on her leash to change her momentum by some amount. Newton's third law says that every force that acts produces an equal and opposite reaction force—so a human trying to stop a running dog by pulling on her leash will feel a force pulling him toward the dog.

Noether's theorem: A mathematical theorem proved in 1915 by the German mathematician Emmy Noether, showing that symmetries in the laws of physics lead to conserved quantities. For example, the fact that the laws of physics do not change in time—that is, that they will work the same way tomorrow as they did yesterday—leads to the conservation of energy.

particle accelerator: A device that accelerates charged subatomic particles to very high speeds, giving them a great deal of kinetic energy, then directs them onto some target. The largest accelerator ever built is the Large Hadron Collider at CERN outside Geneva, which accelerates protons to an energy of 7 TeV (equivalent to accelerating them between two plates with 7 trillion volts between them) and slams them together to convert the kinetic energy into new particles.

principle of relativity: The requirement, introduced by Galileo and strengthened by Einstein, that all the laws of physics must work in the same way for any observer in an inertial frame.

principle of the constancy of the speed of light: The second principle introduced by Einstein in his 1905 papers on special relativity, saying that all observers, regardless of how they are moving, observe the same speed of light: $c = 299{,}792{,}458$ m/s. This is a straightforward application of the principle of relativity—the speed of light is determined from Maxwell's equations, which must be the same for all observers—but the idea of light having a constant speed is sufficiently strange that Einstein made it a separate principle for emphasis.

proton: A positively charged particle found in the nuclei of atoms, consisting of two up quarks and one down quark held together by the strong nuclear interaction.

pulsar: A rapidly rotating neutron star producing intense beams of radiation from its magnetic poles. As the poles spin around, distant observers see regular pulses of radiation, one each time the pole swings across their line of sight.

Pythagorean theorem: A famous relationship between the lengths of the sides of a right triangle, saying that the sum of the squares of the lengths of the sides is equal to the square of the length of the hypotenuse. It's commonly attributed to the ancient Greek mathematician Pythagoras (circa 500 BC) but was probably known earlier than that. There are numerous proofs of the theorem, including one devised by James A. Garfield, the twelfth president of the United States.

quantum electrodynamics (QED): The modern theory of interactions between charged particles and light, independently developed by

Richard Feynman, Julian Schwinger, and Sin-Itiro Tomonaga in the late 1940s. The best-known version is Feynman's approach using diagrams as a calculational shortcut, but all three have their uses.

quarks: Fundamental particles in the Standard Model, which come in six varieties: up, down, strange, charm, bottom, and top. The protons and neutrons making up ordinary matter are themselves made of up and down quarks bound together by the strong nuclear interaction. The other four types are unstable and only seen in particle physics experiments.

rest energy: Energy associated with an object that is not moving through space, given by Einstein's famous equation $E = mc^2$, where m is the mass of the object and c is the speed of light. This energy can be attributed to an object's motion through spacetime: even a sleeping dog is always moving forward through time at a speed c.

Riemannian geometry: The geometry of curved space, worked out in detail by German mathematician Bernhard Riemann, in which parallel lines will eventually either cross or diverge if extended far enough. Einstein's theory of general relativity is expressed entirely in terms of the Riemannian geometry of spacetime, which is bent by the presence of matter.

Schwarzchild radius: The minimum radius a sphere of some mass can have before it forms a black hole. When a black hole forms, the event horizon is located at its Schwarzchild radius.

simultaneity: The notion of events occurring at the same time. In classical physics, which assumes a single universal time, simultaneity is simple, but relativity tells us that two events that happen simultaneously according to a stationary dog will happen at different times according to a moving cat or bunny or any observer moving relative to the dog.

singularity: The infinitesimal point at the center of a black hole, containing all its mass. Explaining what happens at the singularity requires a theory of quantum gravity, combining both general relativity and quantum mechanics, but no working theory of quantum gravity has yet been found.

space-like separation: Separation between events that is greater in space than in time. While an observer moving at just the right speed will see the events happening at the same time, no observer will ever see them happening at the same position in space. Events with space-like separation are outside each other's light cones and thus can never be connected by a chain of cause and effect.

spacetime: The combination of space and time required by relativity, because both the distance in space and the time between two events will be different according to different observers. This is a shining example of why physicists need to hire some English majors to help them come up with cooler names for things.

spacetime diagram (Minkowski diagram): A graphical representation of a series of events in special relativity, plotting distance in space on one axis and distance in time (the time multiplied by the speed of light c) on the other. They can be invaluable tools for understanding what different observers see.

spacetime interval (Δs): The invariant distance between two events in spacetime, given by $\Delta s^2 = x^2 - (ct)^2$

If Δs^2 is positive, the events have space-like separation; if Δs^2 is negative, they have time-like separation. All observers, no matter how they are moving, will agree on the value of the spacetime interval.

Standard Model: The current best theory of particle physics, with six quarks (up, down, strange, charm, bottom, top), six leptons (electron, muon, and tau, plus three neutrinos), and sixteen force-carrying bosons (the photon, W^+, W^-, and Z bosons, plus twelve types of gluons). The Standard Model does not provide a quantum theory of gravity, so it can't be the complete theory of everything, but as of 2011, nobody has solid evidence of physics beyond the Standard Model, though not for lack of trying.

stellar aberration: An apparent shift in the position of stars in the sky as the Earth goes around the sun, which poses a major problem for theories attempting to explain the Michelson-Morley experiment by arguing that the Earth "drags" the aether along with it in its orbit.

string theory: A theory attempting to unify general relativity with the Standard Model of particle physics by describing different types of subatomic particles as lengths of string vibrating in different ways. String theory requires several extra dimensions of space beyond the three we're used to, and while it has proven very fruitful mathematically, it has yet to make any predictions that can be experimentally tested, so its success is still an open question.

strong nuclear interaction: A fundamental interaction between quarks that holds groups of quarks together to make protons and neutrons and holds protons and neutrons together inside the nucleus. The analogue of electric charge for the strong nuclear interaction is "color," which takes three possible values, and the force is carried by gluons, which come in twelve types depending on the colors of the interacting quarks.

synchronization: The process of getting two clocks separated by some distance to read the same time. This is more complicated than most dogs think, and special relativity shows that it is impossible to synchronize clocks for all observers. A collection of clocks properly synchronized according to a dog will appear to be out of synch to a cat that is moving relative to the dog, and vice versa.

tidal force: A stretching force arising because the force of gravity decreases with distance, pulling the inner part of an object more strongly than the outer part. As the name suggests, this force from the moon causes the tides in the Earth's oceans. The size of the tidal force depends on how rapidly the force of gravity is changing at a given position. The tidal forces near a black hole are strong enough to rip objects apart into long narrow strings, a process dubbed "spaghettification," which not even a cat deserves.

time dilation: The slowing of time according to the special theory of relativity. In relativity, the rate at which a clock ticks depends on the speed of the clock relative to the observer. A stationary dog watching a moving cat go by will see the cat's clock taking more time between ticks than an identical clock with the dog, and vice versa. This effect has been confirmed using atomic clocks on satellites, jet planes, and even at speeds of only a few meters per second.

time-like separation: Separation between events that is greater in time than in space. While an observer moving at just the right speed will see the events take place at the same position in space, no observer will ever see them take place at the same time. Time-like–separated events are inside each other's light cones and thus can be connected by a chain of cause and effect.

unified theory: A theory explaining all fundamental physical interactions—electromagnetism, the weak nuclear force, the strong nuclear force, and gravity—as different manifestations of the same interaction. Electrical and magnetic forces were unified in Maxwell's equations, and the electromagnetic and weak forces were unified in the 1970s by Steven Weinberg, Abdus Salam, and Sheldon Glashow, but no current theory successfully unifies all four fundamental forces. Finding such a theory is one of the greatest unsolved problems in theoretical physics.

vacuum energy: Energy associated with empty space, which exists because quantum mechanics requires all particles to have both particle-like and wavelike characteristics. In QED, the vacuum energy occasionally manifests as virtual particle–antiparticle pairs. The vacuum energy is probably the source of the dark energy driving the accelerating expansion of the universe, though the exact connection between them is not understood.

velocity: The rate at which an object changes its position in time. Velocity includes both the speed of the object and the direction in which it is moving, because, as any dog who has headed the wrong way down a one-way street can tell you, keeping track of your direction is very important.

virtual particles: Particles in Feynman diagrams that appear and disappear within the diagram, never escaping to the outside, which are a manifestation of the vacuum energy. These virtual particles cause tiny shifts in the energy of charged particles interacting with electric or magnetic fields, which have been measured to fourteen decimal places. Virtual particles are also the source of the Hawking radiation emitted by black holes.

weak nuclear interaction: One of the four fundamental interactions of the Standard Model of particle physics, which gets its name from the fact that it affects things in the nucleus but is not as powerful as the strong nuclear interaction—and because physicists are terrible at naming things. The weak force is carried by the W^+, W^-, and Z bosons and was unified with the electromagnetic interaction in the 1970s.

worldline: A line in a spacetime diagram representing the trajectory of a particular observer through spacetime. The worldline of a sleeping dog is a straight vertical line as she moves forward through time but does not move in space. The worldline of a beam of light is a line at 45 degrees from the vertical, no matter whose observations are represented in the diagram.

FURTHER READING

Numerous popular books have been written about the theory of relativity over the last century, and readers wanting to learn more have a wealth of options. If you want more detail about how special relativity works, David Mermin's *It's About Time: Understanding Einstein's Relativity* and Tatsu Takeuchi's *An Illustrated Guide to Relativity* are two books that grew out of classes for nonscientists (at Cornell University and Virginia Tech, respectively). Brian Cox and Jeff Forshaw's *Why Does* $E = mc^2$*?* gives a very nice overview, with very little math, of how modern physicists think about special relativity. Robert Oerter's *The Theory of Almost Everything* is an excellent introduction to the Standard Model and the properties of all those particles.

General relativity is considerably more mathematical and intimidating, but two very approachable books on the subject are Clifford Will's *Was Einstein Right?*, which explains some of the many experiments confirming general relativity, and Kip Thorne's *Black Holes and Time Warps*, which provides an exhaustive history of thinking about black holes. Both give a good sense of how the theory works without delving deeply into the geometry of curved space, and Chapters 8, 9, and 10 of this volume would've been almost impossible to write without them.

The application of relativity to cosmology (Chapter 11) is covered in numerous books. Two good overviews are Simon Singh's *The Big Bang*, which covers the historical development of the theory, and Richard Panek's *The 4 Percent Universe*, which provides a detailed history of the discovery of dark matter and dark energy, the mysterious substances making up 96 percent of the universe. Sean Carroll's *From Eternity to Here* explores the nature of time as it relates to the history of the universe and includes some fascinating speculations about what happened before the Big Bang and what will happen after the death of the universe as we know it.

Unifying quantum mechanics and gravity remains an open problem in theoretical physics, but that hasn't stopped people from writing about it. Brian Greene's *The Elegant Universe* is a popular introduction to string theory, while Lisa Randall's *Warped Passages* explores more recent developments regarding the possibility of extra dimensions. Lee Smolin's *Three Roads to Quantum Gravity* introduces some of the nonstring approaches to the problem of reconciling the two most successful theories in the history of science.

Readers interested in the historical development of the theory should enjoy Thorne's *Black Holes and Time Warps* and Singh's *The Big Bang*, mentioned above. Peter Galison's *Einstein's Clocks, Poincaré's Maps* does an excellent job of putting relativity in a historical context and makes an interesting argument that Albert Einstein and Henri Poincaré were both motivated by important technological issues of the day. Einstein himself is a fascinating character and the subject of many biographies, with Walter Isaacson's *Einstein: His Life and Universe* being the most recent and one of the most comprehensive.

The definitive scientific biography of Einstein is Abraham Pais's *Subtle Is the Lord: The Science and the Life of Albert Einstein*, which covers his entire career. It's not for the faint of heart, as Pais doesn't stint on the mathematics, but it provides a fascinating look at the years of faltering steps between special relativity in 1905 and general relativity in 1915.

Finally, if you're interested in a whimsical treatment of the subject (and if you've read all the way to the end of a physics book featuring a talking dog, you must be), George Gamow's Mr. Tompkins stories are a classic, following the dreams of a mild-mannered bank clerk who dozes off during a lecture on relativity. And if comic books are more to your liking, there's also *The Manga Guide to Relativity* by Hideo Nitta, Masafumi Yamamoto, and Keita Takatsu, which even includes a talking dog—though I hasten to add that she's not nearly as cute as Emmy, who remains the best.

INDEX